brilliant

presentation

presentation

third edition

What the best presenters know, do and say

Richard Hall

Prentice Hall
is an imprint of

Harlow, England • London • New York • Boston • San Francisco • Toronto • Sydney • Singapore • Hong Kong
Tokyo • Seoul • Taipei • New Delhi • Cape Town • Madrid • Mexico City • Amsterdam • Munich • Paris • Milan

The right of Richard Hall to be identified as author of this work has been asserted by him in accordance with the Copyright, Designs and Patents Act 1988.

Pearson Education is not responsible for the content of third-party internet sites.

ISBN: 978-0-273-76246-1

British Library Cataloguing-in-Publication Data
A catalogue record for this book is available from the British Library

Library of Congress Cataloging-in-Publication Data
Hall, Richard, 1944-
  Brilliant presentation : what the best presenters know, do and say / Richard Hall. - - 3rd ed.
    p. cm.
  ISBN 978-0-273-76246-1 (pbk.)
  1. Business presentations. 2. Public speaking. I. Title.
  HF5718.22.H35 2011
  658.4'52- -dc23
                    2011030729

The publisher is grateful to the *Financial Times* for permission to reproduce the extract on page 50 from 'Pleasures outweigh the perils of a more balanced commute', 3 July 2006.

10 9 8 7 6 5 4 3 2 1
15 14 13 12 11

Typeset in 10/14pt Plantin Std by 3
Printed in Great Britain by Henry Ling Ltd, at the Dorset Press, Dorchester, Dorset

Contents

About the author

After reading English at Balliol College, Oxford, **Richard Hall** learned all about marketing at Reckitts, Ranks Hovis McDougall and Corgi Toys where he held the position of Sales and Marketing Director. He migrated to advertising as an MD at French Gold Abbott and moved on to become CEO of FCO – the most creative agency around – for 10 years. Following this, he became Deputy Chairman at Euro RSCG. He is currently Chairman of Showcase Presentations, the Friends of St. Nicholas, Brighton and Richard Hall & Associates.

Over the past 17 years Richard has chaired 7 companies, mentored scores of senior executives, written 6 books that have been translated and sold in 22 countries, and moved to Brighton surrounded by family and grandchildren. He is currently learning more than ever before – mainly because we live in such great and changing times.

www.richardhall.biz
richard@hallogram.freeserve.co.uk
http://marketing-creativity-leadership.blogspot.com/

Author's acknowledgements

Brilliant presentation, like brilliant advertising, makes things happen; increases share price and sells product. Quite simply it can make the difference between corporate and career success and failure. And, like all important things, it is undergoing rapid change because people are getting better at it, working harder at it and finding new ways to make an impact. In this third edition of my book I've tried to update you on those changes and make the advice to all would-be-brilliant-presenters up to date, telling and memorable.

My thanks and love to Kate, my wife, who is my most helpful and patient critic. Also thanks to Samantha Jackson, my original commissioning editor and to Eloise Cook who is managing this edition. She and a large team of excellent people at Pearson do the company and themselves credit by their energy, enthusiasm and brilliant support to writers like me.

Introduction

Presentations are as normal as meetings nowadays

The world of work has made the communication of information and ideas to your team or a wider audience normal day-to-day stuff. You're expected to be good at it. The transition from this to a bigger audience needs some coaching, to be sure, but it's all about the same thing. Communication that is clear, compelling and confident.

Since my first and second edition more and more presentations of various kinds are being made and more and more people are finding themselves being asked to make them. I also coach people in the art of presenting and I have been, as it were, midwife to some very good presentations from some very nervous (nervous when they started, that is) presenters over the past three years.

But this means expectations are changing

In sport nothing stands still. Technology in equipment, better diet and training techniques and the simple human urge to raise the bar and beat records, prevail.

So it is with presentations. Over the past years the quality of presentations has improved inexorably and people now are taking more risks: speaking without notes, speaking to bigger audiences and finding themselves performing alongside accomplished professionals.

This book examines those changes and the opportunities these create, as well as firmly laying down the basic skills any communicator needs. There is little point being a skilled slide producer but an inept speaker. Getting your 'toolkit' in shape is your first priority. All good things start at the beginning.

The importance of presentations

Since communicating your ideas and data dominates the world of work and increasingly the world of charity, local community and good deeds, being a good presenter has never been more important. In fact, being a brilliant presenter can transform your career.

 'In an information economy the ability to convey facts and argument clearly may be the most valuable skill of all.'

Thomas Weber, Stanford University

On the other hand being a poor, unprepared or inarticulate presenter can spell doom, however good you are at doing your job in other respects.

 'Incompetent presentation is tantamount to fraud.'

Financial Times

In terms of its scale of importance, Gregory S. Berns (Professor of Neuroeconomics, Director of the Centre for Neuropolicy, and Professor in the Economics Department, Emory University) puts this strongly:

 'A person can have the greatest idea in the world – completely different and novel – but if that person can't convince enough other people it doesn't matter.'

The art of convincing people – persuading them that you are right – is the highest form of oratory and presentation that I can imagine. But suppose one day you have a great idea and need to tell people.

- Will they listen to you?
- Will they believe you, and in it?

It isn't easy – but you can do it

Being a brilliant presenter is not easy. In point of fact being a brilliant anything isn't easy. You need some talent, a huge desire to get better, a practical toolkit, a great deal of time to practise and a lot of hard work and discipline. If becoming a good presenter isn't easy then becoming a brilliant one is going to be very hard for most of us.

This book gives you the basic toolkit, some critical pointers and a lot of words of encouragement. Read it and apply it and you will be a good if not brilliant presenter in a few months. But you must have the right attitude. You must be prepared to be honest about where you stand in the league table of presenters now and you must be happy to put in the hours of practice and homework you will need to help move you up that league table. It should also make you feel relaxed about your quest for excellence and show you that you are starting this journey with the majority of your peers.

It's amazing that so many companies and so many executives assume that presentation is a natural art like walking when it's more like swimming and much more dangerous. Take anyone and throw them in the deep end and watch them swim ... or, more likely, panic. Like swimming, presenting is something you get taught and something you learn. Some people are innately confident and think nothing of standing up in front of a room of people and talking to them. If you are one of those lucky few you still need to read this book because then you might, just might, become a brilliant presenter, a gold medal orator. Many, however, will flail, swallow water, flounder and go under.

A cruel form of torture – the 'P' word

All you have to do is walk into a room and shout 'presentation time' in order to reduce half the people there to quivering wrecks. One person I coached found it hard to remember their own name and what their job was when made to stand up in a room empty of other people apart from me. I could even see their legs quaking, so powerful was their ability to visualise what the presentation event was going to be like. To them, sudden and unexpected death would have been preferable to this torture.

So we have the three issues that impede a brilliant or even a simply competent presentation being delivered:

1 Nerves so bad they create what I call 'panic drowners'.

2 Lack of knowledge as to what to do. This breed is very common and I call these 'non-swimmers'.

3 Lack of preparation and creation of a simple story leading to confused misery – or 'poolside waverers', who belong to the school Warren Buffett describes thus: 'No one knows who isn't wearing trunks until the tide goes out'.

What happens if you can't present well

A former chief executive of Rentokil said memorably after a presentation to analysts that was criticised particularly harshly: 'I'm paid to be a CEO not an actor.' Shortly afterwards, because of analyst reaction to his presentation, he was fired.

Tony Hayward, ex CEO of BP during the Deepwater Horizon crisis of 2010, faced the same problem of being revealed as a poor communicator. He lost his job too.

If you can't present well you're unlikely to win. A judge at a contest for a bursary at a university once told me that from the series of submissions of business ideas the winner was chosen not so much because hers was the best idea but because hers had been the best presentation.

We are, all of us, exposed daily to very competent presentations by people called newsreaders and hosts on chat shows. We have a benchmark. It is no longer acceptable to appear nervous or shifty or lost for words. We must not be what high court judges sometimes describe as an 'unreliable witness'.

And over the past few years this standard of competence has improved thanks to people like Steve Jobs, Sebastian Coe, Morgan Spurlock and academics like David Starkey and Brian Cox.

The so-called 'transparency' of contemporary life in which more managers and company leaders are simply asked more and more questions is one where having the ability to stand up and talk fluently about almost anything and, more importantly, stand up for your company with clarity and command, is the least we all expect.

So competent presentation matters a lot. Investors in companies take a great deal of notice of the confidence and conviction with which a leader talks about his or her company and its plans. How else, they argue, can they tell what the real prospects are? If the chief executive is uncertain about what is going on, or if it seems he or she is probably spinning them a lie, why on earth should they put that precious pension money at risk in his or her care? A senior analyst once told me that he wanted 'to see the whites of their eyes'. In other words, he demanded a brilliant, convincing and coherent presentation told by someone who seemed to be telling the truth.

In saying this he represented the confrontational nature of the analyst–manager relationship. Similarly Jeremy Paxman doesn't need to look into the whites of any politician's eyes because he starts from the assumption that 'they're lying to me and it's up to me to find out about what'.

And if presentation has a slightly dodgy reputation with some people it's because of people like Alistair Campbell and Peter Mandelson, the architects and brilliant presenters of New Labour, and Karl Rove, who was George Bush's Mr Fixit. They helped to create a new level of aggressive spin that made many of us uncertain of what was true, and led to the assumption that all politicians were probably telling porkies.

The 'must-haves' in business today

Away from the increasing murky and spinning-a-line world of politics, however, there's a short list of must-haves if you want to get on in life. Here's the checklist for candidates:

- Can they read?
- Can they write?
- Can they add up?
- Can they perform simple tasks consistently?
- Can they do what they are told?
- Can they present?

Presenting is taught not just as a normal skill at many schools now, but as a key module in the teaching of English alongside reading and writing. Would that we non-swimmers from a previous generation had been as lucky as the children of today.

Presentations are normal nowadays

At almost every level of every company the need to do a presentation of some sort will occur quite often. This could be a simple speech of thanks – that's a presentation – or a description of the activities of your department to a visitor – that's a presentation too – or the story about the new marketing campaign accompanied by visual aids and music – that is a big presentation.

At the very least, regard this book as an insurance policy. Buy it, read it and when the call to do a presentation comes you'll be fine, provided you follow the advice it contains. But if you aspire to being the person who is constantly asked to present stuff because you are very good at it, and you want people to describe you as 'brilliant' not merely 'good', then read, mark, learn, inwardly digest and practise like mad.

 'Practice isn't the thing you do once you are good. It's the thing you do that makes you good.'

Malcolm Gladwell

There are different levels of competence so we are going to go through this step by step and start at the beginning. If that's too basic for you, you can jump to the appropriate level.

Your first step towards being a brilliant presenter has already been taken: you are reading this book, the most up-to-date (and updated), practical and simple 'how-to book' on the subject.

Now all you've got to do is put in the hours. Good luck.

PART 1

Changes in expectations

S ince I wrote the first edition of *Brilliant Presentation* some things have changed and some haven't.

What has changed?

What has changed is the dominant style of presentation from lecturing to something more conversational. From American rouse-the-troops, if you like, to European informal, the presenter today more closely resembles a stand-up entertainer than an under-rehearsed business executive.

What has also changed is the importance of the presentation in determining career success, from firstly getting a job to then being identified as a high flier. Today, presenting is one of the key business attributes. If you can't present convincingly you are unlikely to progress as fast as your other merits might suggest you should.

Finally, what has changed is the competition. More and more people are becoming competent and confident presenters. The standards of performance have risen dramatically and many presenters have become more ambitious and more experimental.

What hasn't changed?

What hasn't changed is the need to devote time and effort in the quest to become good at presenting. Some have natural talent but talent without effort will achieve little or, at any rate, less than it should.

What hasn't changed is the need to prepare and to have a good story, and the ability to bring it to life.

What hasn't changed is the gut-wrenching fear that many have when asked to appear on stage and present to a large, expectant audience.

And what hasn't changed is the need to understand your audience, to sense how they feel, what they expect and what they need.

Even the different forms of presentation remain similar, of which there are four distinct types:

1 **The exposition**: The data-reporting presentation – company results, investor presentations, research debriefs – where clarity is everything and any hint of hype is to be avoided.

2 **The showpiece**: The big announcement. The result of personal experience and intensive work presentation. The finished story. Designed to inspire, inform and engage.

3 **The conversation**: The work-in-progress, creative, we-are-moving-towards discovery piece. 'Here are the ingredients ... and here's what the dish might look like.'

4 **The sales pitch**: This may be a conversation and probably should be. But work-in-progress it is not. It needs to end closing the sale. You will be asking for an order or a positive response to this tender.

The way you approach each type is clearly different. The exposition will require adherence to corporate guidelines, be understated, be visibly comparable with previous such statements and be thorough and clear. This is about money. The key audience will not be bored.

The showpiece is career shaping, really important and akin to a political presentation. You are looking to move hearts as well as minds. You want votes. This is what Sir Stuart Rose did when he arrived at an embattled M&S and produced a tour de force. His successor, Marc Bolland, is more into the exposition school of presenting, letting actions and data do the talking.

The conversation is a relationship builder, an exchange of ideas. Conversations are what drive most business presentations today.

And while a sales pitch can be conversational, it also needs to move strongly towards a conclusion – a yes or a no. It needs to be inspirational, confident and logical but with plenty of opportunity to react there and then to positive or negative responses.

Presenting can change lives

And of course it always could change lives, so there's no change here. However, *more* people have cottoned on to the fact that a great presentation can win an election, a large contract or secure a hefty salary increase. Quite simply, the power of presenting can change lives and destinies.

There's been an improvement in the standard and style of presentations against which people will judge yours, the tools and technology available and the increasing trend to the presentation being seen as a key management tool.

And underlying this is a truth which, if you can master, is a real career-changer:

The art of communication and communication skill has grown vastly in importance because it's the cheapest and most potent way of changing things and making stuff happen. Presenting skill today is what an MBA was a generation ago.

CHAPTER 1

Changes, technology, the media and you

We can underestimate current changes in business or we can exaggerate them. What we can't do is ignore them. Here's what Seth Godin the author of 'Purple Cow' said in one of his blogs recently:

It takes a long time for a generation to come around to significant revolutionary change. The newspaper business, the steel business, law firms, the car business, the record business, even computers … one by one, our industries are being turned upside down, and so quickly that it requires us to change faster than we'd like.

It's unpleasant, it's not fair, but it's all we've got. The sooner we realize that the world has changed, the sooner we can accept it and make something of what we've got. Whining isn't a scalable solution.

Even the world of presenting has turned upside down but, like Picasso, we must learn to draw before we try the more difficult stuff.

Get the basics right first

Let us start by assuming you need help. That's why you bought this book. Let us also assume (I can do this because I've never yet been proved wrong on this one) that you are a whole lot better at presenting than you think you are. But maybe, for whatever reason, you've failed to brush up on and develop your skills. Or perhaps your confidence has taken a knock and you are just a bit scared of presenting.

Action this day

There's a funny trait human beings have in putting off things they know need doing yet they hate because they aren't very good at them or because they are frightened of doing them badly.

Let's assume yours is presenting. And let's assume there's a quick fix, which might be to close your eyes and visualise what doing a well applauded presentation feels like.

In other words, try to see this task from the perspective of reasonably confident success rather than nervous failure. This would, let's face it, be a great start. The art of visualisation is certainly not new but the extent to which it is talked about especially by sportsman is. As long ago as 2001 Leon Kreitzman was telling people to 'dream their presentations' – in other words, try to see in your mind's eye what the thing as a whole might shape up like.

Stop thinking about what they (the audience) might do to you – dislike you, heckle, slow hand clap, walk out – and what, instead, you might do to them – interest, entertain, inspire.

 **tip**

Thinking positively is half the battle in killing thoughts like 'I'm no good at this and never will be'.

How technology helps

I am something of a sceptic when it comes to technology. Too much emphasis seems to be placed on how clever it is, rather than what it can do for you. Until, that is, Apple began to flex its muscles and do interesting and useful stuff. Steve Jobs' philosophy is instructive:

 'You've got to start with the customer experience and work back toward the technology not the other way round.'

The details on technology are dealt with in Chapter 11 but there are a series of developments that really help. These are described below.

Better slides

- **PowerPoint 2010**: The new singing and dancing system preferred by all who use it.
- **Keynote**: The Apple equivalent of PowerPoint and it works fine, not better but fine.
- **Animation/video**: It's easy to have slides with animation or with video built in. In fact, looking as good as *News at Ten* is possible if you try hard enough.

Theatre

- **Radio lapel mikes**: You used to be trapped behind a lectern. Now everyone has the ability to roam about.
- **Lighting**: This has been transformed by technology and helps to make events look theatrically professional rather than what used to be conference-dingy.
- **Big screens**: Showcase recently staged a huge event at the Royal Opera House in full sunlight using a giant screen which looked fabulous – the sort of thing you see action replays on at cricket grounds.

Roving twenty-first century presenter

- **iPad 2**: This little baby lets you control your slides, contains your notes pages, lights your face as you walk and talk and makes you look like a very modern and cool presenter.

Interaction

- **Twitter**: I saw this used at the State of the Arts Conference in March 2011 where 400 delegates recorded Twitter comments on what was being said on stage as it was being said. Thus 'what rubbish this guy is talking' was countered by 'but he has a good if cruel point about Ed Vaizey' and was being accessed by the audience and to those outside the conference room in real time. A real game-changer this, I thought.

- **Webcasting**: Conferences can now appear around the world as they occur. The value of the big event is multiplied.
- **Teleconferencing**: This has now reached new levels of impact, reliability and accessibility.

Presenting commando

This rather crude description refers to the lone presenter on a naked stage with nothing to lean on, no lectern, no slides, no props. Just as technology makes whizz bangs easier to achieve, so a lot of people are going 'back to basics' – check it out and see if it works for you.

It's up to you

Technology can help us do unusual and sophisticated things affordably and well. But in the end how good your presentation comes across is up to you, not technology, techniques or technicians.

> **brilliant** tip
>
> Use modern technology but don't be ruled by it. They came to hear and see you, not your computer.

You have nowhere to hide

The one thing that doesn't change is that sick feeling in the pit of some people's stomachs when told they have to present something. No technology can make that sense of horror dissipate. Something more powerful than technology can overcome it – human ingenuity and the will to succeed.

For those who've been to those schools that now teach communication skills it's less traumatic. For those promoting a book or on a roadshow it becomes a normal and iterative activity. In this lies a simple discovery, which is that most people's horror of presenting lies in the fact it's usually a one-off – like a driving test or a first (and only) night. So if the process could be made to feel less one-off and in consequence less dramatic it might prove easier.

How?

By practising it again and again and again; by preparing for a month ahead; by avoiding the high-risk strategy of a last-minute busk.

A huge piece of advice is to start young. I constantly come across people who've managed to avoid doing a big presentation through cunning, probably that same cunning that got them to the top of the greasy pole. But eventually they always get found out and discover they're on at the IOD Conference at the Albert Hall talking to thousands … or worse.

Every time you get a chance to present, take it and put in the hours so you give it your best shot and learn a lot. Remember what Malcom Gladwell said:

 'Practice isn't the thing you do once you are good. It's the thing you do that makes you good.'

It's so simple – the more you do the better you'll get. So stop hiding and start presenting.

Techniques get rewritten every day – this is theatre

The big presentation in front of hundreds is hard to separate from theatre because this is precisely what it is.

Increasingly we are seeing a new generation of 'stand-up presenters' who stroll on stage and talk fluently for half an hour or so without a note or a slide in sight. Bobby Rao, ex Marketing man at Vodafone and now a founder of Hermes Venture Capital, is the best I've seen at this. Incredibly poised, knowledgeable and taking a lofty overview of everything, he dominates the stage through his command, conviction and certainty. Perhaps certainty worries some of us in such an uncertain world. But I recall the words of that eminent judge Lord Denning whose Pupil Master told him when he was a young man: 'People pay us for our certainties, not for our doubts.'

I was once talking to Adam Crowley who is a master presentation producer and who described what he said was 'wonderful': a presenter sending their rough slides through at 9.30am, arriving at 11am and going on stage at 1.30pm. Was this busking or fresh? I love the idea of 'fresh presentations' in which the material used is totally current and topical. We'll soon see presentations in which presenters at the end of a day cross-reference slides and comments made earlier in the day.

Members of the Magic Circle like Nick Fitzherbert use the rules of magic to maximise audience focus when they present. They approve of Steve Jobs who, like any magician, holds his latest product up there in front of his face.

 **brilliant** tip

If you have a prop, hold it up next to your face.

Techniques used in business schools are creeping into presentations in which the Q&A session starts from the second the presenter walks on stage and the whole content is built around a framework informed in terms of content by audience participation. At its best this works brilliantly.

High theatre uses every technique in the theatrical canon. Trapeze artists, music, video, costume changes, pyrotechnics ... the trouble is we are increasingly into a world of content where the brightest of audiences have been to *Les Miserables* and *War Horse* and don't want stardust sprinkled on the message. The era of the melodramatic product launch may have passed.

But this brings us back to Steve Jobs who is the biggest game-changer in the world of presentations. Carmine Gallo's book on Steve promises to tell us how 'to be insanely great in front of any audience'. Steve is given a very good introduction:

 'Steve Jobs is the most captivating communicator on the world stage. No one else comes close. A Jobs presentation unleashes a rush of dopamine into the brains of his audience.'

I've watched Jobs and he's unquestionably the smartest product presenter and pitcher of innovation I've seen. His tricks are simplicity, suspense and certainty. He makes the audience out there believe in him. And he plays on their desire to hear more. Like my grandson when he begs to be told the story again, Steve is like the storytellers of old, telling of brave Apple and the mighty monster Microsoft.

David Cameron briefly changed the world. Imagine a politician unscripted; a politician who learned his script. Amazing! But the allure has since worn off because the effort required to do this the whole time ate into his busy, powerful life. Now he's back to pedestrian Dave.

 recap

All around, people are getting up there and talking. And many are doing it very well. Presentation is more than a business tool now, it's the primary way of achieving what TED (Technology, Entertainment, Design – the forum for great speakers to do their thing) calls 'ideas worth spreading'. Check it out: **www.ted.com**

What's great is seeing performers like Morgan Spurlock and Matt Ridley putting across interesting and brilliant ideas, when the performance isn't intruding on their thoughtfulness and the use of technology is kept to a minimum.

Technology is great, but let's give three cheers for content.

CHAPTER 2

This is your career we're talking about

ncreasingly it's no use pretending that how you look and how you sound is irrelevant to how you are perceived as doing your job. A scruffy, grumbly and mumbling dentist will do less business than someone who is cheerful and looks well-kempt. There was a dentist in New Barnet when I was young called Dr Screech but that's another story.

These days your competence and potential as an executive will be judged by how well you present. And in a global economy there is no longer anywhere to hide. So even if you are a success at your job, good at the key skills of project management, people management and creativity, if you can't do compelling presentations your career prospects are shackled.

'Can you present?' is top of the job-panels list

It's 9.30 and the interviews for the big job on procurement have started. There are three external and two internal candidates. One of the internal guys is a woman who's very good and very nervous. She's also very tired because she's currently doing three jobs brilliantly – hers, her boss's (who's on holiday) and her assistant's (who's not yet been appointed). You know what happens. One of the untried and unknown external candidates gets the job because 'That was simply the best presentation … brilliant. And frankly Sheila was a disappointment … didn't hack it on the day.' Sheila had been doing a good job for three years, a great job for one year and a slightly flustered presentation for

20 minutes. All those great appraisals were to no avail, it was PowerPoint and body language that did for her.

Whose fault: the panel or Sheila's? Panels, however professional, tend to be very subjective and impressionable. They take against a candidate's tie or their voice. They seldom really study the CV except for gaps. But they tend to love that presentation because it's their chance to, in automotive parlance, 'kick the car's tyres'. So in the end, panels being panels and their expectations being for the exam question to be brilliantly answered, this is the fault of Sheila's holidaying boss and the HR people in failing to help Sheila understand this and prepare properly. But Sheila is also to blame. She hadn't listened to the way the company worked and what it wanted. Which was a dazzling presentation.

 **brilliant** tip

Your presentation in a job interview is a clincher. Don't underestimate its importance.

In a situation in which Sheila had the job sewn up, all she could do was let it slip from her grasp by being careless. A presentation is the easiest way to achieve a baffling defeat or an unforeseen victory. Like exams (where the real skill is to conceal your ignorance and parade your strengths) they are an easy way to mark a candidate there and then. As a panel member you get to see people under pressure and judge their ability to organise their thoughts.

Corporations love presentations

They really do. I've heard people describe themselves as being sucked into their own company's way of doing things. At IBM in the bad old days the way to the top lay in having the best slide deck. Someone told me he went to a company presentation over two days and saw so many presentations that his head ached with bullet points and wonderful slides. There were over 1900 slides, he said, which the organisers strangely seemed to regard as something of a triumph.

Presentations are business's way of creating collective buy-in and information flow – in theory, in theory. Jack Welch, ex CEO of GE and perhaps the most successful and famous business leader in our lifetime, said of creating slides:

'I've always thought that chart-making clarified my thinking better than anything else. Reducing a complex problem to a simple chart excited the hell out of me… I love doing charts and got so much out of them. The crazy thing about it was that we always felt the last presentation was the best ever.'

The story here is that if people like Jack Welch think presentations are that important you had better listen. Especially in tough business environments where senior executives love shining the spotlight on you and asking tough questions just as you are getting into your flow. At one fast-moving consumer goods company you are told you are only as good as your last presentation. This means your working life is a bit like running a non-stop Grand National, and if you fall at Bechers Brook you are toast!

Just as meetings feed the business day, so presentations are the currency by which many places judge their executives.

brilliant tip

Meetings feed the business day and your presentations at them are the way they judge you.

So the best survival tip I can give anyone is to get your presentation skills up to scratch ahead of anything else. In a tough world you can present your way out of trouble but, however smart you are, a bad presentation is a public and unerasable black mark.

How else do you pitch an idea?

Houston, we have a problem. I have a great idea: a career-accelerating idea, an idea that could make us all rich. But I have a problem. My audience:

● has a very low attention span;
● is intolerant of ideas because it hears new ones the whole time;
● has had its fingers burned countless times;
● is minded to give all who present to it a hard time ... just because it happened to them.

And I am not very good at presenting.

So what arc my chances of success?

Well, how I can I put this? Not that good or, to be more precise, probably zero.

Remember what Greg Berns said:

 'A person can have the greatest idea in the world but if that person can't convince enough other people it doesn't matter.'

So it's *how* you present it that makes your idea matter.

brilliant tip

If you have a great idea you'll kill it by presenting it badly. So learn to present brilliantly as well as creating ideas brilliantly. Both matter as much as each other.

The man leapt on to the desk

That man was Kevin Spacey pitching a film idea to the top guys at Pathé Films in London. In fascinated horror they watched as he tap-danced and sang on the CEO's treasured desk just feet away from their embarrassed faces.

I've seen Kevin present and he's one of the very best, helped by an uncanny knack of mimicry of Bill Clinton, Morgan Freeman and Jack Lemmon. He is of course an actor and he nervelessly held the audience in the palm of his hand, talking about why governments and business needed to support the arts and why cutting funding was, to any sane person, not only unthinkable but an act of criminal vandalism.

You are unlikely to be as good as Kevin. He's to presenting what Lee Westwood is to golf or Marcus Wareing is to cooking. But with constant practice you can be very good. Use a great coach and you too can be brilliant.

 **brilliant** tip

Kevin Spacey got brilliant through hard work.

Improving your reputation in your business

Ask to go on a presentation coaching programme. Say that you realise how important it is to be a better presenter. If there's a great presenter in your business, ask if he or she can help you. The key is to go public about wanting to be great at the art of public communication.

In doing this you are staking a claim to be a presenter to keep an eye on. Once you do that you have to keep your side of the promise by proving you are really good. And you'll do this by practice, practice, practice and by working with experts on improving all aspects of what you do.

 **brilliant** recap

Your career is reliant on your ability to be a great communicator, so start getting good by putting in the hours.

If you put a priority on presenting you'll find you create time for it. You'll start rehearsing and thinking about it when you're in a traffic jam, when you're showering or when you're waiting for a call. You can rehearse anywhere if you're keen enough.

Do not leave getting good at presenting too late. It takes so much longer and is so much harder if you wait until you're a senior executive before confronting the fact you don't like it, don't want to do it and aren't any good at it.

And if you are quite good at it, get better – get brilliant, because brilliant presenters are more hirable than mere brilliant operators. Brilliant presenters change minds, sell change and influence the way things develop in a business.

It's this simple – if you want to transform your career, become a brilliant presenter.

CHAPTER 3

Really understanding your audience

You may have all the talent in the world, have practised thoroughly, have a solid story and great slides and, what is more, be feeling good and confident but things can still go wrong. You have to do even more – you have to *really* understand your audience.

The audience usually wants you to succeed

Provided you haven't stood up in front of a bunch of sadists or hardened cynics, your audience would prefer you to be informative, interesting or entertaining. Even if they are pessimists they'd almost certainly like you to surprise them with your brilliance. Give them the benefit of the doubt and try to start by loving them and see if they'll love you. Usually they will.

Why is the audience there? And what's their agenda?

They may have been told to attend, they may have chosen to attend or they may even have paid to attend. You'd better know which and broadly what is on their minds – their expectations, hopes and fears. You need to know what they already know, what they want to know, how simple your presentation need be or how sophisticated they'd like it to be. You need to know who they are and what interests they represent. And what, if anything, they know about you.

brilliant tip

> You don't have good or bad audiences. You have audiences you read or you fail to read.

And you also have to remember that you are, as a presenter, in the business of giving impressions. Brilliant presenters give first and foremost a brilliant impression. They do this by looking confident, friendly and thoroughly at ease with their subject and the people they are talking to. When audiences relax with their speaker something magical can occur. This is best expressed by Maya Angelou:

> 'People will forget what you said, people will forget what you did, but people will never forget how you made them feel.'

brilliant tip

> Learn from the best around. Pause for a second and try to recall how people you've heard recently made you feel.

Brilliance is wrecked by not listening

I recently heard the story about a well-known performer at conferences, known as a 'motivational speaker'. These are often terrifying people who have manic staring eyes and who speak very quickly using a lot of acronyms: 'TEAM – together everyone achieves more'. He arrived late owing to terrible traffic caused by an accident on the A1. Even Mr Motivational Cool was breathing a bit heavily as he rushed into the conference centre for the annual Spartemex conference. His contact from the company rushed up to him: 'Thank God you're here, we were so worried. You're on pretty well right away and they really need cheering up a bit. Everyone's very down because ...' 'No time', the speaker interrupted. 'Sorry. I'm bursting for a pee and I need just two minutes alone to get my head together. See you by the stage.'

Two minutes later he sprang on stage, giving big eye contact and bellowing 'Hello, Spartemex – and how are we today?' Silence. Virtually everyone was looking down. 'Come on, boys and girls, get it together because you are a great team and you know what TEAM means – T ... E ... A ... M – together everyone achieves more.' A girl in the front row stood up, looked at him (at last, a reaction), burst into tears and rushed off down the aisle howling. After half an hour he staggered off stage, having tried everything and got no reaction apart from people shaking their heads and walking out.

He spotted his Spartemex contact. 'What was all that about?' he asked. 'That is the only disaster I've ever had and the worst audience I've ever come across.' His contact shook her head sadly. 'You were in such a state when you arrived that you didn't give me a chance to explain, and then you just pushed past me to get on stage. Our deputy chairman collapsed on stage in the middle of his speech of welcome and died. We all loved him. Everyone's feeling bereft.'

Mr Motivation *had* briefly wondered about the ambulance that nearly drove him off the road as he made his hurried approach to the building. But he was so wound up in his own world he wasn't listening to the world of his audience.

Lessons in how to understand audiences

Always make sure that you are clearly and fully briefed, especially on any areas of sensitivity. The marketing director is Irish, so no Irish jokes; the new chief is a mathematician, so no jokes at all; the deputy chairman has just died on stage, so don't be jolly.

 **tip**

Make sure that you are always clearly and fully briefed, especially on any areas of sensitivity.

However hyped up you are, whatever process you use to get your adrenalin going, do not turn off your brain. Listen, look around and understand what's going on.

When in doubt don't dig yourself into a hole. It's smart to get among the audience and see how they feel. And, by the way, never overestimate your ability to change the mood of a meeting.

Not all, or indeed many, audiences are homogeneous. Understand all the undercurrents. Do not present radical material to dyed-in-the-wool conservatives and expect applause. Don't invest your money on a potentially losing hand when you don't even understand the game you are playing. Get, or try to get, the audience on your side. Don't be boring. Get to the nub of what will 'sell'.

Don't drink or party before a presentation because it slows the brain. If you are presenting as a team then get it together. Never let one player go off dribbling the ball alone. Applause does not mean a good review. Pleasing people with your presentation is not the same as winning with it. You are there to win, not entertain. Listen to where your audience is, indulge them and hit the hot buttons. And if you press a button and nothing happens, move on.

 **example**

A presentation to NBC Super Channel

When I was deputy chairman of the advertising agency Euro RSCG we pitched for the NBC Super Channel business. No one had heard of it in the UK at the time. It now seems to be a cable channel. It's frankly rather confusing. Back in the mid-1990s I discovered one hotel – the Hyde Park as I recall – that was testing it out.

Being an energetic spirit I booked in, ate room service food, drank beer and watched the channel from 7pm till 2am. Now, I never watch much TV except when I have flu so this was a strange experience. I watched people talking about their health – their legs, bowel movements and breathlessness – a session on share prices, some baseball, a chat show with Jay Leno and, as I got sleepier, I seem to recall some soft porn.

In our straightforward presentation, in which boxes were being ticked or not, I had a slot in which to review the product and as time passed it seemed that I was probably the only person in the room (including some of those from NBC) who had an intimate and extensive experience of the product.

When I came to talk about it I must have taken leave of my senses. I extemporised and exaggerated about Jay Leno, about vitamin pills, about the business correspondent's cleavage ('the Dow's well down this evening…'). I even managed to make baseball sound funny.

The guys and girls from NBC were in hysterics. My own team was laughing but in a slightly bemused 'follow that' sort of way. When it came to it, my contribution was very funny but unfocused. Judge a presentation not by the applause but by the reviews – ours were not good and we did not win the business. Never let the performance devour the plot. Upstaging? Eat your heart out, Brad Pitt. The NBC team left thanking me, ignoring the rest of the boring team (yes, they were very boring) and drying their eyes.

brilliant tip

Not only know your audience but also try to know their feelings, needs and hopes… Oh, and be relevant to their needs.

Overall, you should not only know your audience but also know their feelings, needs and hopes. You should respond to the vibes you get – and if you don't feel any, provoke some. You need to work with the minds of people in the audience and develop what is going on in there.

I have often been asked to speak at weddings, funerals, birthdays and other events. The brilliant thing about these is that the audiences are:

- full of politics – families always have undercurrents;
- replete with landmines – don't make any German, Irish, mother-in-law or whatever jokes;
- focused on the event in question;
- friendly – willing you to succeed;
- diverse in age and sensitivity;
- waiting to be entertained or undergo a catharsis.

I've had more satisfaction and learned more about the pure art of presentation doing such events than anything else. Focusing on the subject and making sure that you romance, dramatise and make the subject look good is everything. Being upbeat and optimistic always works – even at funerals. Not 'This is a catastrophic tragedy' so much as 'She died too young, of course, but look at all the great stuff she did while she was alive – and boy, was she alive.'

brilliant tip

Audiences like good news but most of all they like clear, understandable news.

brilliant recap

The audience is always right, so it's your job to understand them, not theirs to understand you. You are the magician so it's up to you to make magic. If you are given the opportunity to do any speeches, celebratory or valedictory, or quite simply any presentations to a real audience, take the opportunity. Giving a speech is a wonderful experience and you only discover the core truths of presenting when you actually do it live.

The core truths are:

- know your audience;
- know your subject;
- write a great, positive, and good natured story;
- simplify it;
- add splashes of colour;
- aim to entertain;
- listen and respond to your audience;
- relax with them;
- 'act', so you achieve maximum effect.

And this is the very best advice of all: love your audience and you'll find they love you.

PART 2

Knowing yourself helps you improve

The seventeenth-century poet John Donne said 'no man is an island' (before gender equality had arrived in town). He was right. We can only exist in a civilised global society by comparing ourselves with our peers and competitors, in our behaviour, our culture and our standards.

Presentation is an activity that takes place everywhere from Beijing to Bexhill, from Delhi to Wellington ... the whole time. So who's doing it well? Who's nervous? Who's adding to the art of presenting? Who's making the leap from presenter to persuader ... and how are they doing it?

Aspiring to nerveless brilliance

In this part of the book I'm going to deal with the two biggest issues inhibiting the performance of many presenters:

1 Nerves – the sort of nerves that stop you thinking straight and prevent presenting being a pleasure; nerves that induce such suffering people can't see how smart you really are.

2 How good you are – not knowing what 'good' is, not knowing how you actually compare with your peers, not understanding how to be smart enough to be (when appropriate) a 'post bullet-point' presenter.

The perspectives that give confidence

You know how it is when you are lying in bed at 3am and everything goes out of proportion. When your capacity to hypothesise is vivid and gloomily creative; when you are incapable of snoozing and focus on losing. This is surely the most negative you'll ever get.

So here are the two key perspectives that will help:

1 Sort out what's going on inside our own heads.

2 Understand what's going on out there in the wider world.

Overall I want you to see just how good you could be, should be and deserve to be at presenting; how to do yourself justice and how to enjoy the experience. Self-knowledge and the ability to benchmark where you stand in terms of presentational competence are big first steps on the journey to brilliance. And if you are thinking, 'Well, it's all very well for him. He doesn't know how I feel and how bad I am', the answer is I do and I do! But below the surface you surely have the ability to improve dramatically ... everyone I coach becomes brilliant. Truthfully. Not in a day or a week but sooner than they could ever imagine.

And to cheer you up here's what Stevie Wonder said. Let it be your motto too:

 'We all have ability. The difference is how we use it.'

CHAPTER 4

Fear – how to understand and control it

Controlling your nerves (and nearly everyone has nerves) is the biggest step you can make in being good at and enjoying presenting. To control them you must understand them and realise that nerves are normal. **You must also realise that controlling them is an absolute must for anyone who aspires to being a brilliant presenter.**

To many people the whole business of doing a presentation, or indeed any solo performance of anything in front of an audience, causes physical anguish so great that it is disabling.

The former chief of a major multinational told me that he'd cried himself to sleep following an inaugural speech he'd made at his new company, which he felt had been ruined by his faltering and then stuttering and drying up. He said that he'd felt like a high-performance Italian car running out of fuel on a busy motorway.

There is little point in talking about the finer points of presenting if the prospective presenter or victim (because this is how they may feel) wants to get it over and done with before bursting into tears or worse.

It's the voice that is key

In Julie Stanford's compendium *The Essential Business Guide*, Chris Davidson gives some great advice. He notes that 'the voice is an excellent barometer of the body's overall level of nervousness'. He recommends the following:

- hold good posture;
- breathe from the diaphragm;
- keep the voice well lubricated with water;
- pronounce whole words;
- relax the face and shoulders (especially the neck and shoulders);
- do a warm-up – especially making '...ng' sounds as in 'bring' or 'bang'.

The last two of these are the really essential ones. If your neck isn't relaxed you're going to struggle to sound or look good. This is great stuff and I endorse it – especially the advice on posture and relaxation, and also the warm-up. Go to any opera and you'll hear them warming up an hour or so ahead. If your voice is so important, and it is, give it the chance of being your ally by giving it some preparatory exercise.

How to avoid freezing

The voice coach Valentine Palmer also focuses on confidence. He talks about 'freezing', 'losing your way' and 'your voice letting you down at a crucial moment'.

A presenter freezing is like an actor drying up on stage. It is not something that anyone who has conquered it wants to talk about because, like blushing, it is hopefully an embarrassing thing of the past.

You avoid freezing by keeping light on your feet, moving around and telling yourself: 'This will be OK.' Cars don't seize up until the oil runs out. Keeping talking to yourself is the equivalent of lubrication. Your reassurance to yourself is the oil of self-confidence. You avoid losing your way by ensuring that you keep signposts and a roadmap in front of you – your script or notes will do that. However, if you are one of those contemporary brave souls who is speaking note-free (the presentational equivalent of 'going commando'), keep the structure very simple: plan a speech in no more than three sections, with no more than three subsections in each section.

Your voice is your best friend or your worst enemy. Feed it Tic Tacs beforehand and make sure that you water it on the stage – frequently. If it seems about to give up in a croak, take a swig and think something like 'Ah, vodka – nothing quite like it.' That might at least make you smile.

 **brilliant** tip

Your voice is your best friend. Look after it with Tic Tacs, water and warm-ups.

I want to dwell on this issue of nerves at some length because as long as you feel a 'confidence crisis' or the threat of a 'panic attack' you will find presenting hard. You may survive an encounter with a presentation if you are very, very nervous but it's unlikely that you'll ever get beyond being graded as 'quite good' until you are in consistent command of yourself and especially until you're in command of your voice.

There are exceptions of course. Arthur Rubinstein used to be sick before going on stage to play the piano. Resistant to the last, he had to be pushed reluctantly on. Then he played the piano and he was, of course, wonderful! Many great actors are quivering wrecks before a performance, while others crack jokes. Nerves express themselves differently in different people.

So how do brilliant presenters actually feel?

There may be a tingle or two and a rush of adrenalin, but none of that wild-eyed, dry-mouthed croaking and freezing that has blighted so many careers. However, I know of no accomplished presenter who cannot take a deep breath and, using any or all of the techniques I'll describe, be in a state of self-control before a performance and *enjoy* going for it when it starts.

I heard of one female executive who lost it so badly that she froze and stared manically at the bemused audience before screaming 'Stop it! Stop it! Just stop looking at me like that ...' before rushing off. Poor thing – she had acute 'presentationitis' which, while not life-threatening, is definitely career-threatening and very, very nasty. So how do we cure this?

Be honest with yourself

You have to start by being totally honest with yourself and others.

- How bad is your problem? Write it down now in detail.
- Describe how really awful you feel when it's at its worst.
- What is the worst it's ever been? No – confess to it, what was really the absolute worst?
- Analyse how you felt – top to toe and top to bottom:
  - how was your mouth?
  - how was your voice?
  - how was your breathing?
  - how were your eyes?
  - how was your stomach?
  - how was your head?
  - could you think clearly?
- And how did you feel about this situation you were in?

 **brilliant** tip

Facing up to your nerves will make a huge difference and stop you hiding from the problem.

What is actually going on?

These acute attacks of nerves have been described and analysed in great detail by squads of experts. Successful treatment is very much the norm, although left untreated such attacks can lead eventually to agoraphobia, which is much more serious. The symptoms include shaking, a racing heart, sweating, difficulty catching breath, chest pain, dizziness, tingling in the hands and nausea, or more extremely (according to *The Analyst*) a palpable, screaming fear rising inside you.

Attacks are more frequent in winter than summer and more common in women than men. Common indications include changes in blood pH levels, which is suggestive of hyperventilation. Apparently, breathing more slowly (or breathing into a paper bag) can restore normal pH levels.

What actually happens? The brain has a small structure called the amygdala which can act as an 'anxiety switch' that flips on only when anxiety seems necessary – when you face a tiger, a raging torrent or the prospect of a presentation. No, I'm not joking – presentations are one of the commonest reasons for the switch being turned on. When it is turned on adrenaline and serotonin are released, making the body – going back to our primate beginnings – ready for escape. That's why adrenaline is also known as the 'fight or flight' hormone. It speeds everything up – the heart beats faster and blood is redirected to the muscles, making you better able to fight or run away. In addition, the brain shuts down – in a life-or-death crisis you need instinct not brains. But in a presentation you need your brain to work, you need to be alert and you need to be confident and look nerveless.

Serotonin is much more complex and in the most extreme situations can kill you (that's a cheery thought – sorry). Think of the effect of magic mushrooms and you'll get the idea. It passes around your system very efficiently and quickly, creating euphoria, over-reaction of the reflexes and a happy drunken state. If you are lucky this makes you feel ready to take on the world, but if you are unlucky it can make you feel as though you are about to have a heart attack.

Essentially your body and brain are undergoing chemical warfare if you get very nervous, and the job of this book and anyone helping a would-be presenter is to make a pre-emptive strike.

So let's do something about it.

Be self-aware

I am neither a doctor nor psychiatrist but, having done hundreds of presentations and watched even more, I know that the most powerful tool you have in preparing for a presentation is self-awareness. If you were flying a plane you'd go through a routine in which you check if everything is working properly. It's called a 'pre-flight check' – a routine, an unfailing, do-it-in-this-order-every-time routine. And this is what *you* must do before you present.

There are a few exercises that may help you, but the chances are you will benefit from working with a professional presentation coach, especially if you are really seriously suffering from nerves to the point of feeling unwell. But do not use drugs, hypnosis or alcohol to obscure the problem. You need to know how to solve it yourself.

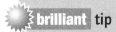 **tip**

Be kind to yourself. Be aware of how you feel. Talk to yourself. Don't be alarmed with your nerves. Work with them.

Do you find the prospect of owning up to it and having the nerves you have treated just a bit humiliating? Well, I guess you had driving lessons? Yes? Well, presenting is like driving but much more dangerous.

To be nervous is normal

The body has the most brilliant defence system and drug drip-feed technology imaginable. You just need to understand what is going on to help to manage the effects and control them.

Everyone gets nervous. Actually, that's not entirely true. Some people *claim* to be free of nerves and, surprise, surprise, they are nearly always hopeless presenters – overbearing, too loud, self-obsessed, and they always over-run.

Feeling nervous is good. Used properly, our nerves are our 'presentation muscles'. Flexing them gives us a 'hyper-performance' feeling that takes us from an ordinary to an extraordinary level of audience control. A sportsperson would describe nerve-harnessing as being key to getting into the zone or the bubble.

The line to be drawn is between:

● 'acutely self-aware and ready to rock' – which is brilliant;
● 'shivering with apprehension and wanting to go to the loo' – which is very bad.

So how do you get to the first state of grace and avoid the second?

As I've said, everyone has nerves. The best, or at any rate the most pragmatic, among us controls them and uses them. Be pragmatic whatever else – the luckiest enjoy their nerves. A bit uncomfortable, but boy, doesn't it make you aware of your own body and your own mortality?

brilliant tips

The most important things are in the mind. Are you giving yourself the best possible chance by massaging your self-esteem? Feel good about yourself, in what you wear, in how your body feels ... freshly bathed ... just exercised. Here are just some of the strategies the best use:

- *Feeling comfortable:* If you feel good you will generally do well. You simply feel in charge of yourself and those around you. Ronald Reagan might have been just a 'B' actor but he was undoubtedly 'A*' when it came to confidence. It was said of Ronald that 'he felt comfortable in his shoes'. Next time you present, make sure you wear your comfiest shoes – and I do mean your comfiest and not your smartest.

- *Practising in the nude:* David Heslop is a former chief of both Mazda and Expotel and is now an entrepreneur/business angel and this is his idea. The idea is that you get a full-length mirror, strip naked and practise your presentation in front of it. Now David is really quite a large man but that merely adds weight to his argument that if you can do something as embarrassing as this with a straight face then a full auditorium will be nothing in comparison.

- *Thinking about your audience:* The certain knowledge of what the audience might think about you may shape your content, your delivery and certainly your confidence. Try the following exercise. Imagine that you have to go into two separate rooms:

a blue room and a red room. You have a presentation to deliver entitled 'How I learned to become a confident presenter'.

- In the blue room is an audience of the most intimidating teachers from school and also your current and all (yes *all*) your previous bosses, and anyone else in your life who has intimidated you.

- In the red room is an audience of relatively well-behaved 11-year-olds whom you don't know, but you do know that they have been told that you are a nice person. Now, I want you to work out exactly how and why your performance will be different in front of the two audiences and why the 11-year-olds may see you differently from the historic ogres. The difference in *your* feelings about each audience may help unlock your inbuilt 'confidence machine'.

● *Preparing thoroughly:* This creates the confidence of the professional presenter. When you watch a brilliant presenter at work and wonder how they do it, you need to remember that they have probably worked much harder at putting it all together than you could imagine. As a rule of thumb, assuming that you are a senior manager planning an important presentation:

- the initial writing should take about 3 hours;

- the refinements about the same – 3 hours;

- your initial slide construction – 2 hours;

- examining the slides you have had done professionally (it's worth it) – 2 hours;

- the rehearsals and rewrites – another 3 hours;

- going back to the drawing board and consulting with colleagues – 3 hours;

- coaching on performance – 3 hours.

▶

● Does 19 hours seem too long? It isn't – ask a professional. What all this preparation does is hugely enhance your confidence and your chances of success.

● *Not being over-dramatic.* You are not going to die. Lucy Kellaway wrote about cycling to work in the *Financial Times* on 3 July 2006: 'Despite the risk I hardly ever feel frightened on my bike. I feel alert and alive but not scared. Recently I was cycling out one hot evening to give a speech to some business people. I was feeling fine about the ride but not about the impending talk. On the way I was nearly hit by a passenger door being flung open, I swerved and narrowly avoided a van. I put the thought to myself "how come I'm not frightened of being crushed to death but I'm terrified of minor humiliation in front of a small audience of civilised people". Suddenly I wasn't frightened any more. On the stage sweaty palms are no longer a problem.' When I had one of those pre-presentation, get-me-out-of-here moments myself I did a little con trick on myself and said 'Imagine I could fly out of myself.' So I did a little mental flight up into the roof beams and looked down at myself. Watching me down there put everything into perspective. I felt quite cheerful – and I performed well.

● *Never drink before you present:* I'm sometimes told a 'shot of vodka' is the answer. It isn't. Don't do it – ever. Any more than you'd expect your surgeon to drink ('just to steady my hands') before an operation. While it may seem the answer for this particular presentation, in your stoned state you won't learn why what went well went well, and why what didn't didn't. Remember Lucky Jim, when Jim got horrendously drunk before a presentation? Never, ever do it.

Brilliant tips on breathing

Learning how to breathe well can make you feel confident and poised. Practise breathing in and out. Take a big breath in and count to four; then breathe out and count to eight. Repeat this four times. Critically, the breaths out are twice as long as those in.

Or sing your nerves away. Anyone who has been unfortunate enough to have had a stutter knows only too well that this affliction happens when talking but never when singing. Try going into the bathroom and belting out a big song, or your presentation message to some kind of rap beat.

And here's the pre-flight check 'breathing clincher' to stop those dreadful 'Why am I doing this?'/'May the world swallow me up' feelings:

- slow down;
- breathe deeply, 4–8–4–8;
- lie down … this can really help;
- practise lip manoeuvres to loosen your mouth muscles;
- visualise yourself being on-form and doing well;
- lie down again – close your eyes and breathe like you've never breathed before;
- do voice exercises to get your vocal cords working;
- make sure your mouth and lips are moist.

 **brilliant** recap

Once conquered, those very nerves that disabled your earlier performances as a presenter can turn you from novice to master. Controlling those nerves and using the tips above will help you learn how to do astonishing and exciting stuff. You need to understand that skating brilliantly involves the prospect of falling, and remember that falling isn't failing – it's learning. It's the same with presenting. Learn as you go.

Anyway, presenting is high-octane dangerous, so you are right to be nervous. At its best it's scary but nice. But whichever way it's always a bit scary. But if you do it well, the post-presentational euphoria is a great sensation. How dangerous is it? Here's what Steven Spielberg said about it: 'We had snakes in *Raiders of the Lost Ark*, bugs in *Indiana Jones and the Temple of Doom*. But supposedly man's greatest fear is public speaking. That'll be in our next picture.'

We're still waiting, Steven. It'll make *Psycho* look like *Mary Poppins*.

CHAPTER 5

The five levels of mastering presenting

This chapter describes the five levels you can achieve as a presenter. As you will see, most people don't get beyond level one. The whole point here is that it's hard to be competent, let alone good, and it's very hard to be brilliant. The key is to be honest about where you think you are starting on this journey of improvement.

This is not an easy journey. You can't suddenly leap up and proclaim 'I want to be a film star' and be one. It will require work, practice, effort and a lot of blood, sweat and tears. Forget the blood – sweat and tears will do.

First of all, let's describe the stages on the journey.

The world of the non-presenter

Before we talk about the levels of mastery themselves, we should agree that there are those who never have and feel that they never could deliver a presentation. These probably represent more than half of the population. This is nothing to be ashamed of, nor is it something to worry about.

Even if you are one of these, something will happen to you once you've taken this book's advice on how to draft and deliver a simple presentation. About half of you will still hate it and find it nerve-racking, although less so than before, but don't worry because you *will* grow to enjoy it. The other half of the previous presentation-haters will get bitten by the presentation bug. You'll experience the serious sense of power as you control a room of people and focus on working an audience. You will scarcely be able to wait before your next presentation or speech. You have entered that most dangerous of phases called 'new' – as in new driver, new golfer, new jogger. You are likely to sweep all before you, terrify passers-by and be alternately full and then devoid of confidence. Your body will become full of adrenaline, serotonin and nerves.

But if you want to progress, here are the steps you need to take and a description of what reaching each level requires.

Five levels to brilliance

I've created these five levels because to improve you have to know where you start. Imagine life without examinations, a career without promotion and a progression up the hierarchy; imagine a world where there was no good and bad, just cheerfully mediocre.

First of all, be aware of where you start. In all the presentation coaching programmes I run, whether it's through modesty or acute self-criticism, few people rate themselves as a good presenter. We are about to change that.

Level one: novice presenter

Surprisingly few people make it beyond this point. I call them 'weekend presenters' – people who do the odd presentation and can cope perfectly competently with small groups. You are good at your job, be it junior brand manager, personnel executive or management accountant. Presentations are neither very important to you, nor are you especially nervous at the prospect of addressing a breakout group.

You tend to 'busk it' and do your own PowerPoint slides, crammed with bullet points, a few hours before your presentation. This is 'a script on the screen', there to help you, the presenter, not the audience. You make a virtue out of the low-key provision of information. At all costs you avoid the risk of what I call 'theatrical performance', because this increases the potential for greater or more conspicuous failure.

If you are asked if you want lights, make-up, autocue and a sound system, or if you are told that you'll be talking to an audience of 100, or that this can be career-shaping, then your nerves will probably start to kick in.

It's important to recognise that people like you are probably pretty good communicators with your peers. But you'd much rather communicate with them in small and relatively informal groups – sitting down rather than standing up.

brilliant tip

Presenting sitting down and standing up are very different. To be brilliant you must be good on your feet. So always rehearse standing up.

It's also important that we don't criticise you for achieving only level one. It's a rule of this book, and life, that you need to try to get better but that it is far better to be a brilliant presenter at this level than a neurotic presenter at level two.

However, there are things you can think about if you are going to improve:

brilliant tips

- As a novice, focus on de-cluttering slides so they are clearer and simpler for the audience to interact with.
- Discard slides if the presentation is to a small group.
- Give the presentation a splash of colour – something to remember.
- Think about exactly what it is you want to achieve – outcomes, not just inputs.
- Think about the audience.

What level-one presenters should do now (you may be beginning to feel that you fall into this category by now) is to review your last three presentations to see if they made sense, whether they could have been better, how you'd do them differently or better, and how you'd cope if you were asked to do a big presentation to senior management at a conference right now.

You may survive perfectly well as a novice but stepping up to the next level won't harm your career prospects.

Level two: apprentice presenter

Getting to the level of apprentice presenter is, in golfing terms, the equivalent of playing off 18 and regularly breaking 90. In cooking terms it's equivalent to being able to do anything that Jamie Oliver throws at you. You are the kind of person for whom dinner parties hold no terrors.

Apprentice presenters are competent. You know how to put together half-decent slides and a well-argued story. You suffer from nerves but you remain in control. You've probably done about 10 presentations, of which about half have been at all-singing, all-dancing sales events and audiences have been generous in their praise of you.

You realise that you should have spent more time on preparation, but you are a very busy person and the thing that gets left till last is always the presentation. You know that your slides are a bit dull and last time you decided to build in a more visual approach, but when that slide showing a herd of gazelle bolting across a plain came up you couldn't remember what it was supposed to signify so you said 'these deer are frightened just like our competitors'. This got a laugh but it made little sense when the next slide that came up said 'the competitive threat – why we should worry'.

Level-two presenters are ambitious and you realise that your prospects can be enhanced by being a bit more adept. There are several things you need to think about.

brilliant tips

- An apprentice should spend a lot more time on preparation.
- Go for a visual approach but with a verbal clue – having 'victim or predator?' on the gazelle slide might have helped the flow (and, by the way, they *are* gazelle not *deer*, that was just careless).
- Script yourself more tightly, especially as you move to more staged events.
- Work on the beginning, the end and the killer central slide if you are to graduate to the next level.
- Lighten up a little … you will stand out if you let your charm and personality shine through.

Level three: craftsman presenter

At this level, presentation skills will affect your career progress. You will be asked to give speeches at events, and others will be glad to know that such a craftsman is speaking at a conference.

It is widely recognised that you are confident and competent, and that you are completely reliable. You are courteous to technicians and you deliver stylish, well-thought-out presentations which reflect how much time, effort and creativity has been spent on them. In fact, you now spend a huge amount of time preparing – probably an increasing amount for each presentation you do. You keep a 'presentation box' at home into which interesting cuttings, pictures or cartoons go. You have a book of great quotations. You have become a management book junkie. Doing presentations has now become a hobby.

Craftsmen presenters promote themselves as industry experts with a view to being candidates for invitations to international events: What's New 2011? in Las Vegas, a think-in for senior managers; The Innovation Forum in St Lucia 2011; New Wave Thinking in the New World in Shanghai 2012; Why Dubai? 2012; After Tom Peters … in Buenos Aires 2012; New Frontiers, New Thinking Ladders in Sydney 2013. Your speeches have been assembled in a short book by Prentice Hall entitled *The Craft of Originality*. Your reputation is made.

But you are still only a level-three craftsman. If you were to see this rating you'd probably say, somewhat indignantly, 'but everyone says I am brilliant'. Maybe, but sometimes a bit dull, perhaps a bit second-hand. A bit – no, not just a bit – *much* too safe, too solidly in your comfort zone. You're very good but you aren't great. You'll get invited but you won't get star billing.

You no longer get a flutter of nerves when you present. You are in perfect control. You are urbane, funny, supremely confident and consistent. But you too have several things you could address to reach the next level.

brilliant tips

- As a craftsman you should carry on presenting – you like doing it and the audience seems to like you.
- Ask yourself a searching question – what are you achieving?
- Reflect on whether or not your company – which is subsidising your activities – deserves a bigger shout in the messages being put out. Are you selling the company hard enough? Or are you just selling *yourself*?
- Try to be more exciting, frightening, dramatic – anything to get out of the safety box you are in.
- Work with others to raise your game – if you were a golfer we'd be wanting to get you down to low single figures … and you can do that, almost certainly.
- Try to do one controversial and challenging presentation that puts you under a bit of pressure.
- Write a book to set up the new controversial positioning.

Level four: star presenter

This is the most dangerous level of all. At their very best, stars are Oscar winners and incomparably talented – at their worst they are just dreadful.

If you are one you'll suffer agonies before any performance as though your stomach is being eaten alive by ferrets. Before you perform you are horrible to everyone around – technicians, friends, colleagues, lovers, wives – everyone. You want to change everything at the last minute – always.

Star presenters give the words 'prima donna' new meaning. You are supremely confident yet inwardly utterly devoid of confidence. You are a mess of extremes – happy, sad, energetic, inert, loving, vindictive, inclusive and divisive. You are very focused on what you want but you have a very short attention span and often forget what you wanted. You have blasts of huge creativity and then get stuck for days. You cry in private and sometimes in public too. You consistently feel sorry for yourself – your emotional age is seven. You are often a chief executive.

What you are blessed with is the rare and magic gift of being able to communicate a vision or personal dream in a language that resonates with the audience. You use beautifully simple language and short sentences. Really short.

Level-four presenters work with the best slide artists – those who can make a dull graph jive, who can animate the description of an operational process to make people's mouths fall open in comprehension and who make audiences want to look at that screen just to enjoy the spectacle of vivid logic unfolding before them.

Level-four presenters are presentational professionals and human relations amateurs – brilliant but impossible. The reason you don't quite make level five is your selfishness and erratic behaviour. If you see yourself in this characterisation I have two things to say:

1 Congratulations, you clearly have huge talent.

2 Now do the following if you want to become truly, consistently brilliant and level five.

brilliant tips

- There's not much you can say to a star except 'calm down'. (If you want be really irritating say 'calm down, dear'.)
- Learn the tone of voice and look that you want to project and take it from there – the late Sir Laurence Olivier used to say once he had got the walk and the clothes right, everything else followed.

- Learn consistency – it will give you a longer life and earn you more friends.
- Use the most important words in management – 'thank you' and 'well done' – more often. The team around you can help to lift your game on your off-days and to make you fly on your good days. They need to be appreciated.
- Think more about your audience than you currently do.
- Stop showing off so much.
- Control your nerves – you are sometimes brilliant now only by the skin of your teeth.
- Keep on presenting. We need you.

Level five: brilliant presenter

Congratulations! Join a cadre of the elite – Steve Jobs, Ken Robinson, Bill Clinton, Sal Khan, Malcolm Gladwell, Sally Gunnell, Seth Godin, Tim Bell, Alan Parker, Matthew Taylor, Deb Roy and Richard Eyres (and, currently, quite a few others) – mostly American and many presenting through TED and the increasingly important speaking circuit.

The idea, of course, is that books like this, which are dedicated to the improvement of presentation skills, will make 'brilliance' a more attainable peak. In 1953 Everest was scaled for the first time while in 2010 over 500 people climbed it.

Like flu, brilliance is catching.

This is what brilliance is

I define brilliance in presentation as comprising:

- a deep, transcendental knowledge of your subject;

- an all-consuming passion for that subject;
- the ability to tell a story in simple language;
- the ability to make a story seem fresh, full of a sense of discovery and very engaging ('Again, Grandpa, tell it again');
- a mastery of pace and control. Level fives are busy people generously bestowing their time on audiences.

It's asking a lot of the brilliant presenter that they always be brilliant, but once they've set the standard we should expect no less of them. And sometimes a star, or even craftsman, presenter can attain brilliance. Take Melvyn Bragg (a good craftsman presenter I'd thought). I saw him once on stage in Brighton talking about his book *12 Books that Changed the World.* He was a revelation, fizzing with enthusiasm and boyish excitement. He was hopelessly lacking in discipline and ran over time, but he spoke from the depths of knowledge and discovery that drive him. It was like hearing Eric Clapton playing a long improvised riff – totally brilliant.

Brilliant presenters are shameless

Brilliant presenters should polish their craft and coach those around them. There is too little constructive talk about how to create great presentations, yet this is what often elevates debate about a subject and the advancement of human understanding.

Brilliant presenters always know when a risk is worth taking and how to lift and move an audience. Brilliant presenters are shameless. They would lie and cheat to get that buzz back from the audience. That smell of attentive astonishment and the creaking of people sitting on the edge of their seats.

Being a brilliant presenter may be a more achievable attainment for you than you had ever thought possible. Nine out of 10 presenters fail to get better because they don't really try and they can't be bothered. Think about it. Could you be one of these wretches?

Don't be a wretch. Be shameless. Be brilliant.

 **tip**

Be shameless – seduce that audience – don't play safe.

 **recap**

It's possible you won't reach the peak of brilliance as a presenter but it has to be worth trying. And as that legendary advertising man Leo Burnett said: 'If you reach for the stars at least you won't end up with a handful of dust.'

The most important advice is to be ambitious; to want to get really good at it and in short, to go for it.

I asked Will Arnold Baker, Managing Director at advertising agency Publicis, what he thought constituted brilliance in presenting. He said that some of the best he'd seen were utterly shameless (that word again), that to them this was 'show-time' and that they'd say anything and make up anything to achieve a good effect. To them, applause was all that mattered.

PART 3

The presenter's toolkit

The would-be presenters have calmed their nerves, and having established their lack of expertise at presenting are pessimistically resigned to doing this presentation. What follows may allow them to become better than resigned and actually look forward to the experience. Until now they have been looking at the metaphorical piece of wood grimly realising they don't know what a chisel, plane, screwdriver or hammer actually are, let alone how to use them.

This is the toolkit which will enable you to fashion your presentation and, after a bit of practice, learn how to produce a thoughtful, effective and admired piece of work. Brilliance, which is what we are after, is going to be achieved by hard work. So welcome to the workshop; welcome to presentation creation.

The key tools are as follows:

- Understanding the **context** of your presentation – how to maximise its relevance.
- Learning to create **a story** as opposed to a 'deck' of disconnected slides.
- Making the story **colourful** so it engages people with its evidence, drama and illustration.
- Creating **great slides** that multiply the impact of what you say rather than getting in the way.
- **Performing** with power, passion and conviction so the audience is enthralled by you.

- Making sure that it's **your content not your style** they remember.

The acid test – did you get your key message over?

These tools are the ways in which you'll carve, hone, develop and polish your story. But remember they are tools, the means to get your story from your brain into the minds of your audience. How you will be judged is by how effective you are in getting your message across, not by how skillful you are with your tools.

Every presentation has impact at two levels. You need to aim to score well at both but remember the art of persuasion is a long game:

1 First impressions – and these are very important. These will be measured in applause and instant feedback.

2 In the cold light of day. Very often it'll be the presentation with the better and clearer content that will, in the long run, score best.

Good luck with the toolkit and the creation of brilliant presentations.

CHAPTER 6

The context of the presentation

f you don't know *why* you are doing a presentation, *where* you are doing it, *when* you are doing it, precisely to *whom* you are doing it, *what* the state of the political or commercial climate is outside and inside your own company, *what* the audience knows already and *what* they expect and hope from you, then expect to fail – unless of course you are very, very lucky. Understanding the context in which your presentation will take place is the single biggest factor in determining whether or not you can become a brilliant presenter, because there's one simple rule of life: if you don't know what's going on around you, you are unlikely to be successful. Understand the exact reason for a presentation, the reason why it's happening, the desires and needs of the audience, their take on the local and world headlines, the company's latest news, gossip within the company (never underestimate this) and you may find that what you had feared was a rather mediocre piece of work can go down a storm because it is so relevant.

brilliant tip

It's a simple rule of life that if you don't understand what's going on around you, you won't be successful. So pay attention.

The effect of context is mighty. What may be a fantastic presentation on day one may sink like a stone on day two. That all-important audience reaction is always about context. So how do you avoid getting it wrong or how, on a more positive note, do you make sure you get it right? How do you make sure you are always relevant to your audience?

First, get the exam question right

This seems pretty obvious, but is your presentation actually on the subject you've been asked to talk about? Auberon Waugh was apparently ready, at a price, to talk about anything, anytime and anywhere in the world. However, over the phone he misheard the organiser of a particular conference. All he really heard was the word 'Tuscany' followed shortly by the words 'pounds' and 'thousands'. He was, to be sure, somewhat startled to be asked to talk about 'breast feeding'. Nonetheless, Tuscany is Tuscany and thousands of pounds are thousands of pounds so he conscientiously researched the subject about which until then he knew nothing. It was only in Florence as he was about to deliver his piece, that he discovered he was supposed to be talking about press freedom. Now it's an easy mistake to make, especially if you are slightly deaf and rather avaricious.

brilliant tip

Make sure you answer the exam question. Relevance is the key.

It's astonishing how often this 'exam question problem' is got wrong. Here is a set of planning questions you should ask, and have answered, if you're going to optimise your chances of being on brief, doing yourself justice and being brilliant.

↗ **brilliant** checklist

1   What is the event about?

- Why is it happening?
- Who decided to set it up?
- What is the agenda?
- Is there a hidden agenda?
- What is the organiser hoping to achieve?
- What is the parent company doing currently?
- What issues is the company facing commercially, competitively, strategically, financially, politically, in staff-turnover terms, in terms of management stability? Are there many new brooms or old brooms, and how about bad news, good news or scandals?

2   What about the people in the audience?

- Do they like each other?
- What happened last time this group got together? Did they stick around afterwards?
- Was the event a success or a failure?
- Are there any underlying tensions? While running one off-site presentation I found the atmosphere very strange. It transpired that the chief executive and his deputy were at loggerheads, so much so that one (or both) would probably have to leave the business. Their colleagues were taking sides. No one had thought to mention this in briefing me.
- What do the people in the audience do? Are they your subordinates, peers or superiors?
- Does anyone know what they think, what their problems are and what they expect from you and this meeting?
- Do they know you or of you? If so, what does this knowledge comprise? If not, what is their expectation likely to be?   ▶

- Can you get to meet any of them before you speak? David Heslop (ex-chief executive of Mazda Cars and Expotel) was asked to speak to the sales people from the *Independent* newspaper. He prepared his presentation and then realised that it was egocentric, all about what he thought, about his industry not theirs, and that if he did get through to them at all then it would be purely coincidental. He happens to be a good pianist so he arrived a day early and played piano in the hotel bar where, incognito, he got to know many of the *Independent* people. When he spoke next day he did so knowledgeably and as an insider. He was, he told me, a 'wow'.

3   What size of audience?

- How many will there be in the audience? This is critical – there are broadly four kinds of audience size:

   a. An intimate meeting (10 or so): a sit-down, relatively informal environment with, hopefully, plenty of interaction. A flip chart or overhead projector can work well, but avoid elaborate and expensive visual aids.

   b. A big meeting (up to say 30): this is a reasonably formal meeting with fairly strict rules of engagement, which nonetheless will need plenty of back and forth conversation if it is to achieve high energy. This can be a brilliant forum for achieving big realignment of company thinking or strategy.

   c. A properly staged event (around say 100): now we are beginning to talk 'theatre'. You need a stage, good lighting, sound systems and decent visuals. You'll need to have a carefully crafted speech. You may be less likely to improvise and you may need to be more careful.

   d. Real theatre (100+): at this level we are talking about all the tricks of the trade. You need professional back-up, and if you don't have that in spades then you aren't being professional. Spend at least as much time on rehearsing in situ as you would rehearsing in total for a smaller event. This level of performance is high risk and high reward. It's high risk because it is clearly so important, expensive and meaningful. Someone from Unilever told me about 'the good old days' at the Palladium when top management was expected to perform alongside the likes of Tommy Cooper and Eric Morecambe at big sales presentations. Now that really is challenging stuff. And you know what? I bet Eric and Ernie got nervous too.

- A big 'show' is also a big test of you as a manager. If you don't know exactly how that theatre or that stage is going to *feel* then you are running as big a risk as someone going on a journey to a place they've never been before without a map.

4 Where are things heading?

- Think really hard about the present and the future. We are living in a world of profound change. As Leonard Riggio of Barnes & Noble put it in *Fast Company*, 'Everything is in play.' This means that no presentation made to a management group can have much credibility unless it reflects the scale and pace of current change. Every presentation contains the often unstated premise that this is not going to be comfortable, that nothing can be taken for granted and that this event is going to be about disruption.

- There are many good examples of this. Google is talking about its developments being constrained by the speed of light. The size of China is put into context when I tell you that its disabled population is the same as the total UK population. McDonald's finds it quicker, cheaper and more efficient to outsource its customer ordering in its drive-throughs than to do it on the premises. Hedge funds are now so big that collectively they will, in effect, shortly run the world economy. (Or not. The economic crises of the late noughties might have put paid to them.)

- All of which goes to show that there is no certainty in anything nowadays. Consumers are killing marketing as we knew it because by now they all know how it works. Most of the technology we rely on today will be redundant by 2014. Many big corporations are doomed although most will take a long time to die. And if you don't know how to use the web you'll soon expire, Mr Dinosaur.

- So are you fully up to speed with the context of the markets in which you operate? You need to be because no one wants to listen to anyone who is out of date. Business speakers need to be visionaries not historians.

5   Are you up to speed on the day itself?

- By this I mean with what is going on in the news in the wider world, not just in your business. I was once doing an off-site presentation to a strategy group. As I drove to the meeting I heard the dire news of Britain's exit from the European Exchange Rate Mechanism in what was dubbed 'Black Wednesday'. With events unfolding, Norman *'je ne regrette rien'* Lamont, addressed the media and spoke of soaring interest rates. The poor souls at my presentation, having been locked away in their offices, knew nothing of this until I told them. From that point on nothing else mattered to them. They all sat quietly and miserably considering their newly unaffordable mortgages. Context is everything for everyone, always.

6   Where is the event happening and what is on the agenda?

- In the office, close by, luxury hotel, overnight, abroad, big or small location?

- What sort of room?

- What sort of equipment?

- Do you have any say in what can be changed?

- Do you trust the people running it?

- Will it be hot or cold? Whatever it is, make sure you are comfortable and confident – a tight suit and tie in 35°C is a bad idea. I once gave a presentation to a global advertising conference in the Hotel Martinez in Cannes. It was very hot. As I lay sweltering at 2am the night before in my room I abandoned my carefully prepared presentation. I focused on how I looked and the impression I gave. I went 'mafia'. Dark glasses. Black suit. Open-necked, white linen shirt. Rumpled, bedroom hair. They loved the creative look. I talked creatively ... few words, lots of asides, lots of audience contact. I spoke what we now call 'bloke' language. We connected. It worked. I was relevant to the context of the event.

- Who comes before and after you? How is the audience likely to be feeling as you come on? Will they be bored, sleepy, excited, argumentative, angry, fed up, happy, expectant, relieved? Whatever it is, 'play' to the mood and don't fight it. Context is everything and the audience is always right. Remember you can only work with what you've got. At a TED event in London, Mrs Moneypenny (the *FT* journalist) came on right after we'd just seen a powerful and emotive video of Jill Baker describing what it was like having a stroke. 'It was ghastly following that' she told me. But she did brilliantly by acknowledging Jill's story and slowing down her own start so we could all adjust our mood. Audiences can't go from loud to soft or slow to fast in seconds. Turn up the volume and pace carefully.

7   What is the mood on the day?

- The worst mistake any of us can make is to judge everything to precision except, as actors would put it, 'the smell of the audience on the night'. Before you start presenting you must judge how people are feeling and thinking. Are they edgy and unresponsive – maybe they've

▶

had a bad car journey or perhaps a star team player has had a row before leaving home? Sense it and give them all time to settle. You are not the star in isolation. You are the facilitator of their feelings, the organiser of their energy and the conductor of their orchestral mood.

- Remember that perfect performance is not about perfection. Everything is relative to what the audience wants, expects and gives back to the performer. What a coup it was for Neville Coghill and for Oxford to get Richard Burton at the peak of his career to come up to the university and play Dr Faustus. It was at the beginning of his great affair with Elizabeth Taylor so there was a double coup as she was billed to play Helen of Troy. Richard Burton was no stranger to drink and he drank a lot beneath the dreaming spires. Indeed he was rather drunk much of the time and he didn't really know his lines. Now anyone else would have collapsed or been booed off but there was something really odd going on: he really, really knew the meaning, the core essence of the play, its real story. It was undoubtedly the definitive *Dr Faustus*. He delivered its truth, its real emotional meaning and if the details got blurred, well, we can forgive gods like that. Don't get drunk but equally don't confuse perfect preparation with perfect understanding and perfect performance. The detail matters less than the big thrust, the core meaning, the story.

8 How do *you* feel?

- Nerves – you've not entirely conquered them but you have controlled them. You are really quite commanding when presenting. But suppose you feel ill – you are sweating, you have an upset stomach, you feel as if you want to die (if you are a man then you probably are dying). If only the cruel audience knew just how you felt. But how you feel physically is not the issue. Are you focused and alert and 'up for it'? That's what matters. Keep how you feel to yourself. As far as the audience is concerned they need to believe you always feel wonderful.

9 Do you know what the audience feels?

- Get a feel for what it's actually like being in the audience by going and sitting where the audience will sit and getting someone else to talk from where you'll be talking.

10 Are you really in charge?

- With those management skills you obviously possess you can encourage, cajole and tell the people who do coffee, tea and lunch to work with you rather than against you. So do that.

- Get there early and have chats with the people who are often ignored. Whatever happens, at least they will be on your side when your presentation starts. A great presentation to a small group was sabotaged (slightly but badly) by the presenter having a silly and undignified disagreement with a waitress who subsequently insisted on pointing at him and loudly describing him as 'that idiot over there'.

brilliant recap

If you don't know where you are or why you're there, you can't really know what you are doing. Nor can you do it – whatever it is – particularly well. More importantly, what you say will lack that sharp edge of relevance all brilliant presenters have. I believe that the skill of a brilliant presenter is to know that they've covered the following:

- Why is it happening?
- Where is it happening?
- When is it happening?

- Who will be listening?
- What do they want?
- What do they expect?
- What do they need?
- What's been happening in the world?
- What's been happening in their lives?
- What are their biggest problems right now, today?

It's hard to stress quite how important getting the 'context' thing right is. Yet most of us are a bit lazy at doing the research and having the conversations ahead of time that help us understand the background factors which could reduce a potentially brilliant presentation to failure.

Most of all, remember the presentation is now ... not in the past or the future, but right now. What happens now is relevant to the audience and your 'getting' this is the magic key to you understanding how to make the theatre of the event work for you as opposed to against you. Be current, be relevant and you'll reach your audience.

CHAPTER 7

How to write a brilliant story

A brilliant story

One of the winners of the UK Whitbread prize for literature is an American, James Shapiro, who claims that he hated Shakespeare at school but is now besotted with him. His book is called *1599* and tells the story of that one year in which William Shakespeare wrote, among other things, *Henry V*, *Julius Caesar*, *As You Like It* and *Hamlet* – not bad for one year's output, not bad stories.

brilliant tip

Be like Shakespeare: a really great storyteller.

Give audiences that narrative drive Shakespeare created: a strong beginning, a coherent middle and a powerful, surprising and memorable ending.

brilliant tip

Content is king. Getting your story right takes time but unless you can learn how to create an interesting story you won't ever be a good presenter.

Death to mission statements and other jargon

In one of the businesses that I used to chair I banned the use of that ghastly phrase 'mission statement' and insisted it was replaced by 'our story'. Actions like this will encourage people away from corporate speak, jargon and bullet points, towards simplicity, excitement and common sense. And incidentally I loved the advertisement for Ben & Jerry's in which they gently derided the whole idea of corporate jargon. The ad said 'Mission Statement: To Make Nice Ice Cream.' That's wonderful.

The great thing about stories is, after all, that they have something to reveal, something to explain, something to resolve. In other words, something interesting to say.

 **tip**

Never use more than three points in creating a presentation. Think three – beginning, middle, end. All good things come in threes.

brilliant dos and don'ts

Do

✔ Limit your thinking to the 'rule of three'. This simple technique means you are never allowed to use more than three points. Never. Try it and see how powerful it is. For example, there are just three things to say about creating a story: stories are designed to command the attention of the audience; stories are about people and events; and all stories are about change.

✔ Try to summarise the thrust of your story in just a few words. For example, if you say 'This presentation is about our growth, how we've achieved it and what we are going to do next', this will 'fix' in the audience's and your mind the journey you are going to take. Many of us find it hard to say exactly where we are going, which probably accounts for the fact that so few of us actually get anywhere.

✔ Try to create a simple structure so that the presentational story has an order to it. Something like: 1 This is where we were; 2 This is where we are now; 3 This is where we want to be.

✔ Give your audience the setting for the story in graphic terms. Define the time, place, temperature, ambience and so on because they hook into people's imaginations. Do you remember those stories that started 'It was a dark and stormy night'? Already the audience is tingling...

Don't

✔ Start work at the beginning. Start by deciding what the 'end' is, i.e. what the final point is that you want to make. Everything you do must lead up to this final point – everything.

✔ Be squeamish about cutting out anything in your presentation which doesn't drive the story on, however nice you think it is.

✔ Only speak in the third person. Try to introduce first-person experiences and anecdotes. For example, I remember a powerful story I told the Department of Health when I was in advertising and we pitched for a teenage anti-smoking campaign. It was about my niece who was given cigarettes by her mother when she went to parties so that if she felt that she really had to smoke her mother knew she wasn't at risk of being given one laced with skunk. This anecdote spoke volumes to ministers and their officials.

A way of thinking about your story is to regard it as the framework around which everything else fits. It's a simple truth that if you haven't got a good, sound story to tell then you haven't got a good presentation.

Truth is a powerful weapon

We are all impressed by what we rather admiringly call 'true stories'. These are narratives that sound as though they might be fictional – 'you're never going to believe this ...' – but are grounded in reality and actually happened. Well-written historical fiction carries enormous credibility because it's framed in facts that we all know. So with your own presentations, insert facts, figures, names and so on which establish them as being based on truth.

> ### brilliant tip
>
> Use personal anecdote to make this your presentation not a third-hand story.

Anecdote wins hands down in a contest with data when it comes to presentations. The politician who personalises his story with an account of a real event grabs the headlines. Presentations in business will be based, almost by definition, on truth (unless of course you are Enron or one of those corporations that get wrapped up in their own hype). How do you turn a liturgy of figures and mission statements (oh dear – not them) into a narrative, a story that commands attention, inspires belief and, hopefully, takes the business forward?

The advice I've given already should help, but if you want to achieve brilliance then you have to go further than simply applying a mechanical process. You have to learn how that inter-action between story, storyteller and audience works.

> ### brilliant tip
>
> Make your story sound fresh and new – as if it has only just happened, and also as though it has just happened to you.

Does the story grab the imagination?

Because children are the most fearless of critics and the most expert in establishing whether a story is good or bad, try your story out on a child. And since the harshest critics of all are stroppy teenagers, if you really want to stress-test a story to see if it's boring or irrelevant ask a teenager what they think.

 **tip**

Brilliant question. Could an intelligent 14 year old follow your argument? Could they, even they did follow it, be bothered to? If not, you're in trouble.

In the golden, youthful era of storytelling, the acid test of a good story is whether or not it inspires – the *Harry Potter* phenomenon is proof enough of that. In our sophisticated, high-tech world, tales of low-level witchcraft have dazzled a generation or two. It has also shown how many adults have 'childlike' minds and are none the worse for that either. When the *Harry Potter* novels came out they went to the top of the bestseller lists in France and Germany – not the French or German versions (published about eight months later) but in English. Picture the poor young folks in Marseille or Munich reading these 600-page tomes slow word by slow word just to get their heads round that story.

Become a story student

Listen to the stories that people tell at work, at home, in the pub – wherever. Work out why some work and some fail. Why some are brilliant and others not. Listen to comedians and watch the news. Become a 'story addict'. Learn what seems to work and then adopt the role of storyteller and try to grip the attention of your audience through the simplicity, conviction and narrative drive of your story.

 **tip**

Grip the attention of your audience through the simplicity, conviction and narrative drive of your story.

And practise, practise, practise. It takes time to get a good story that is compelling, but keep at it because it's worth it. And don't forget to try it on a child or a teenager. Set yourself a 'story test'. Keep on asking yourself if the story you are telling contains the 'brilliant story elements'.

The elements of a brilliant story

It needs to be:

- interesting;
- relevant;
- surprising;
- involving;
- full of action and reaction;
- well structured;
- memorable; and
- make demands on your audience – well, you do want them to do something as a result of it, don't you?

The best presentations are brilliant stories

Some of the best presentations I've heard have been stories of discovery and learning. Others have been sermons – one of the great art forms of all time where a 10-minute story should be designed to make you laugh and frown, and think and remember on different levels.

One of the best I ever heard came from Tony O'Reilly. He played the Irish British Lion card brilliantly, talking fluently about rugby, the heavy-drinking Brendan Behan and mixing this adeptly with insights into the grocery trade.

I loved the probably apocryphal story of him at Heinz as a very young man determined to win brownie points for his keenness from a workaholic boss. He'd get in really early – say 6.30am – and be there head down when his boss passed by shortly after 7am. Having established his presence he crept off to a quiet part of the building for an hour's snooze at about 7.15am. That's style. It's also quite sensible.

Great memory hook, but what's the story?

One of the most interesting presentations I ever went to was at M&C Saatchi's Golden Square offices where the School of Economic Science put on an event.

Graham Fink, creative director of M&C Saatchi, spoke first and spoke well, without notes but rather nervously. He stressed the importance of creativity, asserted that he was very creative (which he is) and showed some ads. Meanwhile there was a middle-aged lady with her hair done up in a scarf sitting on stage knitting and ignoring what was going on around her. She turned out to be Fink's mother. The reason she was there was to capture our attention. She was charming but her presence was only relevant as a memory hook. I concluded memory hooks alone are not enough.

Telling it like it is – brilliantly

The late Michael Mayne, former Dean of Westminster, was heavily scripted when he spoke. He read his script – each word – but every word was pregnant with meaning. He was brilliant at describing the story of how transcendental experiences could be obtained, in virtually equal measure, from great art, great music, great poetry and from spiritual experiences themselves. The whole experience was magical, a cornucopia of pleasures heard sitting at the feet of a genius.

How stories have changed

Audiences have less time now. Many books are long, like Ken Follett's great stories, but in general terms the long presentation has been replaced by a more sound-bite style event. TED has had a huge influence on this, with speakers learning to work without notes and tell highly personal stories about things that have shaped, changed or transformed their lives. Ken Robinson could obviously talk forever in his easy, conversational and inclusive style about the subject he knows and loves.

Even straightforward sales pitches or reviews of sales performance can be told with more drama and given interesting twists. There's no need to be boring.

 **brilliant** tip

Talk about what you know and do so interestingly and with drama.

Someone who knows and loves their subject is the key to the new world of presenting; someone who is clearly an expert and passionate about telling the story of their journey. We have moved from the more objective world of the recent past to a more subjective one: Steve Jobs talking at Stanford University about his life and his thoughts on death after his liver transplant; Jill Baker on her near-death experience having a stroke – and then, just as we are being gripped by this harrowing story, she gives an extra devastating twist 'And you know I kept thinking – this is so cool, because of course I am a brain surgeon.'

The voice of success

People pay a lot of money or travel a long way to hear Bill Clinton, Warren Buffett, Steve Jobs, Tony Blair, Jack Welch, Malcolm Gladwell or Henry Kissinger. Because they know their stuff. They've been there, they know what's what. They have an overview. While we can't all be world statesmen or bestselling journalists, we can earn recognition for being exceedingly well informed in key areas.

Earn the reputation of having points of view worth listening to and this will make audience anticipation something worth working with. Do not ever underestimate the power of reputation. Because we know that someone who really knows about what's going on will always tell a better story.

 **brilliant** tip

Earn a reputation for specific knowledge and develop it.

The voice of learning

The voices of the good and skilled academics are always worth hearing – the best at IMD in Switzerland or the London Business School are voices that have spoken with and observed the most powerful, the biggest and the most vulnerable. These are the diagnosticians of success and the morticians to failure. They are nearly always worth listening to, with memories full of anecdotes, learning and analyses.

The way they tend to see the change in things is, on a day-to-day basis, less to do with storytelling and more to do with 'sharing ideas'. But to wider audiences, through academics becoming media stars, their stage has now become a place where the mysteries of the universe, the human body or the history of man have become epic stories and popular TV. David Starkey,

Brian Cox and Lady Susan Greenfield have brought science and history to what some call dismissively 'the latte society'. And you know what? They are all great presenters and great storytellers because they make the unfathomable easy to understand without patronising their audience.

 **tip**

Learn how to make the difficult to understand easy to grasp without talking down to people.

The voice of a winner

The great Olympiads Colin Jackson, Sally Gunnell, Amy Williams, Leon Taylor, Steve Backley, Roger Black, Greg Searle and others have all become great storytellers, people who pitch that unique experience and endeavour with finesse, drama and humour, knowing of course that they have a great punchline for their conclusion – the medal winning performance.

It's always that battle against the odds: no money, no resources, injury, loss of form or a collapse in confidence. It's the big man or woman made little man or woman by circumstance fighting back and winning. It's rather like a Western.

It's also inspiring. That formula of talent, pain, failure, fear, anticipation and success by the lead actor in that drama makes for a brilliant story if told as well as these guys do. Amy Williams, the British skeleton racer, telling the story of preparation, including the one we wanted to ask but didn't – 'What if you want to pee when you're all wrapped up at the top of the mountain?' Answer – you spend years on bladder training so you have perfect self-control when it matters.

And I love the story of the young Colin Jackson rushing to his hero Daley Thompson's room to show him his silver medal only to be told 'Oh, I didn't know they did them in that colour, son.' Priceless.

 **brilliant** recap

Most of us are much better at telling stories than we think. These stories don't have to be original (none of Shakespeare's were) but they have to be good, solid, structured and told well. Storytellers give themselves time and space and have good material. Take the Bible, which is full of great stories. Take Milton's *Paradise Lost*, a great story that is very exciting too. Or Philip Pullman's Dark Materials series which is Milton reincarnated.

Another great storyteller is Tom Peters, whose power of delivery may be a greater factor in his success than his ability to dissect a management situation. Yet he tells power story after power story compellingly. I also love his own, personal story. Imagine, he said, your own gravestone. What he wants on his is 'Tom Peters – he was a player'. What a perfect epitaph; what a great climax to a story.

I heard him in Amsterdam some time ago and I thought he was great. He made brilliant use of anecdote and of his vivid experiences in companies such as Southwest Airlines. It was, incidentally, Herb Kelleher, the one-time Chairman of Southwest Airlines, who uttered the down-to-earth immortal line, devoid of jargon: 'Yeah, we have a strategy here – we do things.'

Remember, next time you present, if you haven't got a story then you haven't got a presentation. Whether you're sharing an idea, product, sales situation, business initiative or piece of creativity, tell a story. Be dull at your peril. The world has learned to expect more.

CHAPTER 8

How to give your story colour

There are more and more good presenters out there today. Those presenters are earning promotion and better jobs. They are developing their careers while they are on stage. While many people may have mastered most aspects of presenting, real competitive edge can only be gained by that unusual ability to give your story and your presentation a splash of distinctive colour. The extra plus that makes the audience sit up and really remember what you are saying.

A nearly brilliant presentation in the making

Suppose you have a good story that you've stress-tested by making sure the logic is sound, that it is well structured and even your teenage cousin accepts isn't totally boring. It obeys all the technical tenets of presenting:

- It's relevant to your audience.
- It has a clear message.
- You've applied the 'less is more' rule so ruthlessly that it is now incredibly cryptic.
- It is 10 minutes long even though they asked you to give them half an hour.
- It has no detours, no irrelevance, no shades of anything but … grey.

In short, it has a great framework and is likely to be as dull as tap water – totally clear with just a faint back taste (well, this presentation sounds as though it has been through about 30 audiences already). Anyone could do this – even a Dalek.

Actually a Dalek would probably do it better.

A bare presentation is like bare walls

This chapter is about helping to put the intellectual ornaments and rhetorical colouring into your simple story – the stuff that will turn it from being a plotline into a narrative that engages the audience's mind. In other words, something that has colour, perspective and emphasis. Something with highlights and lowlights with emotional intelligence as well as solid logic, which has a certainty of where it is heading, what its ultimate destination will be but which has surprises and interesting detours en route as well.

Talking about colour reminds me of contemporary classical composer and conductor Eric Whitacre. Eric looks like a rock star. Actually he wanted to be a rock star. At university they said 'Join the choir.' He said 'No, I want to be a rock star.' And then they said 'There are lots of hot women in the choir and there's an expenses-paid trip to Mexico coming up.' So Eric joined the choir. He stood there among the basses regretting his decision until they burst into the Kyrie in Mozart's Requiem. He said: 'It was as though technicolour had suddenly filled my black and white world. That moment changed everything.'

That's what splashes of colour can do.

> ### brilliant tip
>
> Colour, atmosphere and suspense is the stuff that transforms a bare plotline into a narrative that engages the audience's mind.

So how do you find these embellishments? As Dylan Thomas said, 'Let's start at the beginning.' At breakfast time.

Work at breakfast time – be fresh

Read as many papers as you can. Focus on the *Financial Times*, *The Times* and *The Sun*. But always read *The Week* and *Private Eye* too. Snip out or record any good stories that you find. One of the things that freshens any story is something contemporary, something that's just happened, something recent that demonstrates a point you want to make. Be ruthless in choosing a story that makes you look up to date: best of all, of course, is the statement of the type 'This appeared in today's *FT* – what do you make of it? Let me make just three points about it...'

There's something about 'morning fresh' that is captivating. It positions you as an on-the-ball person. It also gives what you say an authoritative perspective and places you in the role of commentator, not just presenter.

A 24/7 sense of curiosity

Don't take anything for granted. A sense of curiosity makes a business presentation sing. Curiosity delivers quirky insights, it is the lead-in to discovering the unusual, the story or fact that can captivate an audience. Curiosity's currency is the line made famous by Michael Caine: 'And not many people know that.' What a sense of curiosity means is keeping your eyes open and seeing what's changed, what's stayed the same, what's new and what's been relaunched.

 **brilliant** tip

Become a question machine and an accumulator of news stories.

Look. Listen. Question. Like Andy Stefanovitch. He's a founder of a US consultancy called Prophet. He talks about the need for a 'museum-mentality', that frame of mind that is awestruck by new, iconic discoveries. He claims to take off a month a year with his family, not on the beach but in New York City people-watching and soaking up art and theatre. He talks of keeping himself filled up and taking it all in. He is Mr Curiosity personified.

Be like Andy who says 'experience everything to empathise with everyone'. Inspiring stuff.

And inspiring stuff is what audiences need

I was recently asked by a client how real audience involvement could be achieved at a relatively small seminar. When I said 'Well, you could start by giving them sweets' I think she thought I'd gone mad. Of course, what I was suggesting was interaction. Sweets, chocolates, anything like that would do – even sampling drinks if it's relevant. Sampling an ordinary Bordeaux and a great New World wine both selling at the same price is a great way of starting a presentation or a debate on the new order and how things are changing today. To make it really interesting you could include Great Wall wine from China too – old world, new world, next world. (Now that's a scary presentation which I'd really like to attend.) Give them ice cream if it's hot and then lead a debate on climate change. Do your presentation but encourage interruptions and questions as you go along.

Have an 'as you present-Twitter commentary' like the RSA did recently. They've also introduced RSA Animate which takes slides to a new dimension – check it out. Take risks, be imaginative, run polls on issues, show the results on screen. It creates amazing tensions and excitements and, for some reason, we all seem to love scores.

Always involve the audience. Do live interviews – get people up on stage with you or pass microphones around the audience.

Think about the role of presenter differently. Matthew Taylor, CEO of the RSA, is breaking new ground as a leading-edge facilitator and speaker. Be like him.

Become a compere not just a presenter.

↗ brilliant checklist

- **Look for the big thought or quote – and keep one eye open for a big idea**: Read the most brilliant management and thought provoking books around. As an aspiring 'big-time-brilliant presenter' you have no choice but to be an avid consumer of the key, current management thinkers – from Steve Levitt to Malcolm Gladwell, from Thomas Friedman to Matt Ridley, from Jim Collins to Sean Meehan, through to a much longer list. These are people who will give you insights, thoughts or who will simply give you amazing slides. You don't have to read all these books from cover to cover of course – just learn to be a great scanner.

- **There is nothing more exciting than to be taken on a visually adventurous presentational journey**: Keep a look out for big pictures, the big changes and trends. These are all around us but we seldom either think or have the courage to use them – pictures of sport, of natural disasters or man-made disruption or anything which has that 'wow' factor, the list is endless. Some will be corny and some contemporary. Some will be pictures you've seen before – from Munch's *Scream* to Picasso's *Guernica*, Hirst's sheep, Turner's sunsets or Tracy Emin's bed. Look out for great news pictures, especially recent ones that will resonate with your audience. Also look out for visual treats in the papers – study the *Financial Times* which, at its best, does some awesome graphic design work with financial results. ▶

- **In a digital age an ideas notebook still wins:** Keep a 'brilliant ideas' book beside you in which the great quotes and insights you gather are kept. Not because you are a trainspotter who has to write everything down but because you never know when they may be useful in fleshing out a presentation. For example, if you are talking about the importance of speed in business today, Mario Andretti, the one-time great Formula 1 driver, said: 'If you're in control you aren't going fast enough.' Now that's similar to the comment made to the British ice skating champion Robin Cousins who was told that he wasn't going to make it internationally because 'he didn't skate to fall'. Consider both of these quotes and what they mean. Could you do a presentation just focusing on these two quotes? Or suppose there's been a corporate marriage. I love this one, by Anthony Hilton in the *Evening Standard* writing on takeovers and mergers: 'The lunch comes immediately; the bill comes later.'

- **Bear in mind that too many words on screen can slow you down:** Don't have dull slides. Jack Welch tells stories of how he and his colleagues spent hours trying to refine just one chart that encapsulated a whole strategic thought. Finding a way of talking about, say, a new people strategy may be as arid as the Sahara when expressed in bullet points but in contrasting the two shapes or, if your people are really clever, morphing from a triangle (normal structure) to, say, a circle (twenty-first century structure) may say more things more quickly and more clearly than you can imagine. Besides which, interesting material and visual material can be more interesting for you to talk to.

- **The next will be printed on an inflatable – what a great way of doing an agenda for a global conference:** Don't underestimate the importance of agendas. They are the key to expectation yet ironically most people regard the agenda as a necessary extra that sits rather drably in the 'conference housekeeping drawer'. But how would you feel about a restaurant with a scruffy menu? Your agenda is your menu. Treat it as your way of previewing how the meeting is going to be – serious and data-filled or fun and innovative, about people or about business plans, about the future or a review of the past. Your agenda's the dustcover to the book you are going to launch, so treat it with great respect and produce something that looks well thought out, attention-getting and important. At recent meetings I've had laminated agendas, pocket-sized agendas, bookmark agendas – anything that takes you from the prosaic to the land of the possible.

● **A good takeaway puts the gloss on a great presentation:** Don't forget takeaways. These are reminders of the event – at analysts' meetings you need presentation printouts on which they can make their copious notes. But these represent a small part of all the meetings and presentations that most people have to do. What every presentation needs is an elegant follow-up, and I love what Martin Conradi at Showcase invented – he calls them 'lunch-books'. They are called this because they are ring-bound, A5 and easy to handle, so you can use them while having lunch without knocking the water or anything else over. They contain the slides and as many appendices as are necessary.

● **Look to have a moment in your presentation that you are looking forward to presenting – a moment to remember:** Always remember the power of the 'money shot'. There is that moment in every Bond film that has people sitting on the edge of their seats – the one that everyone tells their friends about: 'A so-so movie, but that bit where they go over the glacier on a snowmobile is incredible.' In the same way with presentations we are looking for charismatically memorable effects. Now these are not *always* appropriate, especially with small audiences where pyrotechnics will seem superfluous. But for larger audiences you need not only that killer slide but also that moment of drama – maybe borrowed from a favourite film. Maybe it's animation. I recently did a fairly corny but very effective 'tipping point' animation featuring the client using a see-saw that was incredibly hard to move until – whoops – suddenly it tips and all else follows. Corny but memorable.

These are the big thoughts and ideas on colouring-in your presentation, on giving it energy and life. But I also want you and your team to think about how you can become more exciting and how you can keep on trying to achieve brilliant effects.

Be fresh. Be creative. Be spontaneous. Be nice

- **Be fresh:** Think of every presentation as a new one, not one that is re-hashed. A presentation will nearly always be enhanced by individualised splashes of colour because it will appear to be 'freshly cooked', as opposed to being ready-prepared in some presentation factory or, worse still, a congealed leftover from a previous meeting.

- **Be creative:** Think creatively about your presentation. We live in a world of increasing sophistication. Most of us are processing more and more arcane pieces of information. They used to say that information is power. This may be true for Google, but for the rest of us clarity of thought and creativity are what are really powerful. Information is merely there to be used. So use it – use it to intrigue and to educate; use it to make your presentations more fun and more lively.

brilliant tip

Facts sell, facts sizzle. Be specific: time, place, temperature.

- **Be spontaneous:** Thinking on your feet is colourful in itself. One of the most exciting presentations I recently saw, one crammed with colour, was by Matthew Taylor down in Brighton when he talked about 'the white frostbite of austerity' and suggested that the gap between Britain's desire for its future and its current trajectory could only be closed by a richer understanding of human nature. Matthew talks with clarity but ostensibly off the cuff. He creates a wonderful sense of spontaneity. Or Andy Stefanovitch again, recalling a student-created flier he'd found in Starbucks. It said 'Free – blank strips of paper. Useful for a 100 things. Bookmarks. Roll-ups. Small bandages. Decorations, etc.' It made him laugh so much he wanted to hire that kid. Free blank strips of paper. Brilliant spontaneous performance.

brilliant tip

Clarity of thought is really powerful: clarity of thinking on your feet even more so.

- **Be nice:** Guru Margaret Heffernan with panel members like Kirstin Furber (BBC) at a conference on women and why the future is female were smart and they were nice. QED. Their smiles and charm lent colour and diffidence just where it was needed. Pastel splashes of colour. Margaret reflected that today 'nice is the new mean'. Let's hope she's right.

brilliant tip

Do not aim to be good enough because good enough is no good. Colourful, memorable and brilliant is the target.

brilliant recap

The moral is to ask the following questions of yourself and your team at that first critical meeting, when you are writing the brief for your presentation:

- Can we remember that, whatever we are doing, this is going to be brilliant?

- How do we make this presentation different, one that is quite different from expectation?
- How do we give it a real extra wow factor somewhere?
- How do we get audiences to say: 'Very good story, very interesting and memorable too'?

Remember that colour usually comes from the ability to describe and make you feel as though you'd been there yourself (wherever 'there' is). That's why Leon Taylor, Olympic silver medallist diver, describing the pain of diving backwards into a pool from the height of two and a half double-decker buses and reaching 40mph at impact, as 'really hurting', resonates so much. Taylor provides colour, impact and personality.

CHAPTER 9

How to illustrate your presentation

Opera is a narrative brought to life by music. There is a powerful simplicity to many operatic stories but it's the music that gives them emotional complexity and makes them fly. Great visual aids, a little like the music in opera, can take your rhetoric and your argument to a higher level than if you simply rely on the spoken word (though opera lovers will probably shudder at this comparison). The one thing opera is not is understated. Nor should you be as a presenter. Those TED talks are all presentational arias and that's why they work. It's the same with blogs. In today's world we're into a short-form of engaging people.

Think of yourself as the focal point

Even if you don't actually have any slides, don't for one moment imagine that you are doing an un-illustrated presentation – if you have no visual aids, whether slides or other tools of presentation like video or tangible items, you have made a decision to make *you yourself* the illustration. How you look, your body language, what you wear and the way you behave will have even more importance than normal.

And as to whether you really need slides depends on whether you feel more comfortable with them and (critically) whether they will add to the story and make it easier to understand and engage with. This is a 'you decide and only you can' moment.

brilliant tip

You are the focal point, the person they've come to see. Don't let them down by being scruffy, surly and charmless.

Bad slides slow you down

Curiously, some people seem to regard slides as a mere adjunct to a presentation – a kind of necessary evil, a last-minute add-on ('Hey, Mozart, have you got the music?'). It's most strange. Good and sharp visuals can help to drive a presentation, making it easier to understand and more compelling than if there were none, but poor visuals slow everything down; a bit like trying to drive with the handbrake on.

brilliant tip

Poor visuals slow everything down like trying to drive with the handbrake on.

Equally, while anyone can cobble together some half decent, conventional kind of slides, many are full of words and fettered by bullet points – the presentational equivalent of sleeping policemen ... Driving with the handbrake on over sleeping policemen, how much worse can this get?

Professional slides help

On the other hand, a really skilled operator can, with experience, make a PowerPoint presentation look wonderful. I used to work in advertising and there was one piece of advice I consistently gave clients: 'Don't try to do your own advertising.' As with DIY surgery, the result will not necessarily be what you would wish.

By the same token, make sure that you get trained and experienced people to create or at the very least finish off your visuals if you want them to have great impact. Don't try to do them yourself – unless, of course, the casual, amateur look is something you want to project.

Professorial vagueness has its place

Interestingly, many academics seem to be happiest with scratched, out of order and apparently frequently changed overheads. These create a very public denunciation of style. By remaining in beta there's an engaging work-in-progress feel to the event. So far as they are concerned they are saying that substance and unfinished contemplation is all that matters.

Do we imagine that Newton would have had a very snazzy presentation with gorgeous shots of apples? Or would it have been on an old envelope? Or like those of the erratic genius called Robin Hankey, whom I knew at Oxford and who wrote his essays from west to east and then turned the paper round 90 degrees and carried on writing north to south over the previous text?

brilliant tip

It pays to look and sound as though you are thinking about what you are presenting and would be happy to take questions.

We need to look at the context of a presentation if we want to decide on the most appropriate visual techniques and the amount of money we want to spend. But there is a lot to be said for looking as though you are thinking as opposed to reciting a pre-prepared presentation – especially if the subject is very topical.

Less is more

In presentations which are not constrained by lawyers, bankers and analysts, certain rules apply and need restating. I advocate a 'less is more' approach. Keep the number of slides down to perhaps one a minute and go for maximum focus and few words. Few words are key. Talk around a slide or at a slide, never over a slide or reading out a slide. This leads to your slides becoming 'script on screen', which is the ultimate audience nightmare.

Consider replacing words with pictures – but be careful. An account director at a major marketing agency described, with the horror that only an impending disaster can evoke, a presentation where the presenter said something like:

'Last year looked difficult from the outset, though we hadn't realised how rough, especially for some clients and competitors, but we aim to have a smoother passage this year with some great results and happy clients.'

The slides used are inserted in capitals:

'Last year looked difficult from the outset [STORM CLOUD SLIDE], though we hadn't realised how rough [ROUGH SEA SLIDE], especially for some clients and competitors [SHIPWRECK SLIDE], but we aim to have a smoother passage this year [CALM SEA SLIDE] with some great results and happy clients [APPLAUDING CROWD ON HARBOUR SLIDE].'

It was, she said, horrible [ACCOUNT DIRECTOR BEING SICK SLIDE]. No, sorry. Forget that last slide, that is me being silly and applauding because this illustrates so well how things can go wrong.

Consider animation. Include video if you can – for instance, talking about a retail outlet while watching a speeded-up video of it. This can work much more powerfully than looking at a sales graph. Consider big, bold, single word slides such as:

FOCUS

You have to concede it gets your attention.

A favourite idea of mine involved a presentation for a firm of accountants when we wanted to talk about the benefits of partnership – how working together, client and accountant, could achieve synergistic benefits. This was the slide that did that:

$$2 + 2 = 5$$

Being incredibly flexible

Increasingly as we see, conversations are becoming the norm. Kevin Eyres, European MD of LinkedIn, thinks business planning as we used to do it and as it's still done by large corporates is preposterous. Things are moving too fast to write one-, three- or five-year plans. Margaret Heffernan, the expert on women at work, thinks these are things that Alpha males do for comfort. The impact of this trend shows up in the presentation model which is becoming more conversational and kite flying. And, however long it takes to prepare a presentation, be prepared to change it just before the event if there's a change of thinking or events. Remember what J. M. Keynes said: 'If circumstances change, I change my mind.

What do you do?'

Where words and pictures work best together

Mike Weekes is an NLP expert. He's also a world-famous rock climber. At a big conference on 'Performance' at the Lord's Cricket Ground conference complex he spoke about working with Jack Osbourne and getting him back in shape. He also spoke of climbing. The illustration stays with me now.

We see a slide of Mike negotiating an overhang – lying flat upside down with hands and feet connected to rock... That image stays in your brain as you stare in horror at the slide. He tells the story of him, a mere 19 year old in Australia, of his having a hangover (from weed and lager), of his doing this difficult climb with joy in its danger...

What happened then?

One foot slips...

The other foot slips...

And then a hand slips...

So Mike's left holding on by one hand ... above a 400 foot drop...

('I'm there with you, Mike, I'm there and I'm very, very scared', I think)

He says in a matter-of-fact way that he thought was going to die...

('You are Mike' I thought, 'if you are normal.')

And a voice in his head said...'

'How do you know? How do you know?'

It was, he said, like the voice of God.

By then his arm was shaking with lactic acid and his last grip was slipping...

So he swung up his other arm and grabbed...

And got a grip...

And recovered a second handhold.

The rest is history.

There was just one slide up but there was a whole film and a funeral playing in my vertigo-ridden head while Mike spoke and this presentation vividly stays with me ... and those words **'How do you know?'**

Product presentation

The master of product presentation is Steve Jobs. Holding the product up close to his face, he talks about it and praises it. He then has his team pass samples around ... and shuts up.

Product demonstrations are an art. And the best ones are no more than 2 minutes long overall and within them no element lasts more than 20 seconds. They focus on one or two messages, make telling, simple, competitive comparisons and make a big point: 'This is the best in the world because...'

Steve shows things on slide, of course, but he uses props too.

 **tip**

Use props. Hold up your product ... lovingly.

How to use props

Use props: things you can pick up, throw, bounce, balance, examine, hand out. Props force you to relax and the audience to enjoy you. Use the whole breadth of the stage. Present or talk while someone skilled is doing something else relevant while you describe your key point: an expert table tennis player showing how alike table tennis and the internet are – like Marco Montemagno does; someone juggling while you're talking about the complexity of management; dividing your audience into two groups of those agreeing or disagreeing with a proposition and letting debate begin ... Anything is fair so long as it illustrates your point.

Not everything visual has to appear on a screen. Assuming it is relevant, you might put a packet of jelly beans on the seats in the auditorium, or a specially created agenda. Nick Horswell, ex-colleague and founder of the media company PHD, once did a presentation which climaxed (I'm not entirely sure that is the right word) in his walking off stage trouserless from behind the lectern having walked on stage fully trousered. I'm not sure how he did it, undoing his pants and still giving a decent presentation, but he did. Suffice it to say that trousers-down Horswell also brought the house down. His point? I imagine he was saying that just because you couldn't see something, it would be rash to assume it wasn't actually happening.

Don't be trapped and one-dimensional. Nothing excites an audience more than magic ... the magician getting another rabbit or whatever out of a hat. Every presenter is a magician being called on to reveal something new and surprising. So think about ways of revealing your punchline other than by just another slide.

⟋ brilliant checklist

- **The pictures have to illustrate the story you are telling, not something else:** Be clear about what you are saying, to whom and what you want to achieve. Showing pictures of Stormtroopers with the slogan 'Let's go get 'em' may not be that helpful in a presentation about partnership and stakeholder relationships.

- **Your slides need more impact the bigger the hall:** Establish, first of all, how many there are in the audience, how big the hall is, how smart and attuned an audience you have and whether there are language difficulties.

- **Don't be too corporate. Be you:** I have a personal antipathy to corporate templates, which nearly always swamp the presenter's good intentions in 'corporate porridge'. Yet I am a lifelong advocate of brand values. I am a passionate fan of Heinz, with whom I've worked on many projects over the years. You don't need to slam 'HJ Heinz' at the bottom of every slide with bullets shaped like the famous keystone to make it look like a Heinz presentation. Indeed, the best presentation I ever saw from them was done smartly without such constraints and yet it breathed that sense of 'There's no taste like Heinz' or 'It has to be Heinz' in a way that a more mechanistic offering could never have done. If you are not doing a corporate but a more personal presentation, decide in a focused, almost feminine, way on your 'look'. This involves font, colours, feel and style. There is a later chapter in this book that warns against using exotic and mysterious fonts in case in the transportation of a presentation from one computer to another the font is not recognised. I discovered such a font recently which, for obvious reasons, looked good to me on my computer – it is called 'Poor Richard' – but here's how it appeared on another PC: ✎🕮■𝑒𝑟☺○✕■!

Avoid indulgences such as 'Poor Richard' like the plague – life is hard enough without being a presentational punk.

- **If you think of this as hard work and boring it will show:** Enjoy doing your slides. Think of things from the audience's point of view. Think about how to get them on your side, about colour, about simple points. Can you get pictures of any of your audience on screen? Or of the buyer of your product at Tesco? Something that speaks to your audience.

- **The time spent finding the right picture is time well spent:** Don't forget – a picture is worth 10,000 words. But it has to be the right picture. If a presenter is talking about, say, productivity and a picture of a carrot appears on the screen then the audience is likely to be baffled. I recently did a presentation in China which, given the language issues, had to be predominantly visual. The key point is, of course, that finding great visuals that begin to tell a story is very, very time consuming. I and three researchers spent an indecently long time scouring the web to find exactly what I wanted.

- **Sometimes you have to go for it:** Do go for the tour de force. This is the sort of thing the ever-brilliant Richard Eyres, now a non-executive director at the Guardian Media Group, has used in presentations. I've seen him do a great presentation in which he talked, among other ▶

things, about the creation of the Capital Radio website (it had to be 'not leading edge or even cutting edge – it had to be bleeding edge' according to the designers). An executive from Pearson, as I recall hearing it told to me, did a very energetic presentation with a new slide every seven seconds or so. 'Don't do it' begged the producers. But he did. And I'm told it was a complete wow.

- **Variation of look and tone gets people's attention:** What can look like a neat, clear presentation as a booklet to be taken away can be the equivalent of a monotonic drone presented on the big screen. Presentation is theatre … with pauses … with lows and highs, fast bits and slooowwweeerr bits. The visual impact of your presentation can catapult you to the heights or, if it's boring, act like the deadweight of a sea anchor. Don't make it all the same.

- **Briefing your slide designer will reveal to you what you can and can't do and whether you have actually got your act together:** Do compile a brief in cryptic words and pictures. Your designer can't really be expected to read your script and immediately understand what you are trying to achieve. So get a few A5 sheets and create a series of simplistic charts with a thick felt tip pen. This is a trick Showcase, the presentation people, use to stop you putting too many words on a slide. No more than 10 words to a slide is ideal – the best creative directors in advertising used to claim no poster should have more than 6 words, so check some of today's posters. See how you get on being this reductive. Then try it again seeing if you can spot opportunities for some visual fun. Then again to check that the slides tell your story. If you need special emphasis then indicate where this is. Now you are ready to brief your presentation designer.

Why avoiding slides can be wrong

There are many reasons why so many presenters are nervous of slides.

- Most people are used to verbal and written communication, not visual communication (except when they were under five).

- Most people wouldn't know a good piece of visual communication if it bit them on the leg.

- The moment something goes on a slide, technology gets involved and people worry that anything can happen.

- It isn't just technology, it's about being in control. Some people say: 'I'd rather have average PowerPoint slides I've done myself that I can change at the last minute.' I know what they mean but they don't really, really mean it? Do they? Not really. Because being average is not what an aspiring brilliant presenter wants to be.

brilliant example

A guy at Royal Bank of Scotland had rehearsed his big set piece presentation, which was being performed at a foreign location. He'd have scored A+ for preparation. But he went up on stage and the people running the show managed to let their computer go down. Blank screen – crisis! Fortunately he was able to chill out and stay calm. They rebooted and off he went again – 'Whew!' Do your knees feel weak? Mine do. I hate technology, except when it goes right, when I really love it.

It's all a bit like former US president Gerald Ford, of whom it was said he couldn't walk and chew gum at the same time. You have a lot to do and a lot to remember – you have your nerves to conquer, your words to say, things to remember and slides to pay attention to – something's going to crash. Don't laugh at that poor president, empathise with him.

Possibly the worst presentation (there are so many, OK, but one of the worst I ever saw) was when I saw a nervous ad man go off at speed pressing his own slide button … the wrong way. The final slide came up first and his 'Welcome' slide came up last. The slides and his words bore no relation to each other. Having been told never to look behind you at the screen, he didn't, and so had no idea what was going on. There were quiet mutterings among the audience, and then laughter which he thought was at a joke and this just encouraged him the more. He speeded up – he shouted 'Sales will grow' to the accompaniment of a half-dressed girl lying on the bonnet of a car with the headline 'Big Engines are a Turn On'. People in the audience stood up and shouted. He just got louder and faster. Someone approached the stage and tried to help but the presenter resisted, exclaiming 'I shall have my say.'

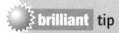 **brilliant** tip

Work with professionals to ensure your slides do you justice.

 **brilliant** recap

You are never allowed to be boring because life is too short for boring presentations, boring ads or boring people. We live in a post-boring world, one in which stand-up presentations are the norm. The standards are rising. What was considered to be a brilliant presentation 20 years ago is probably merely good by today's standards.

Here are the must-haves in presentation illustration:

● have a few very good slides;

● use as few words as possible;

● include quotes (these are always good);

● enjoy your slides and props – they are friends, not hindrances.

Because, apart from anything else, the one element that has really come on over the years is visual design. All kinds of tricks are possible in this wonderful age of special effects. Understand that your slides complement your voice and if you really seek brilliance, you must pay attention to them and their production. Let the 'visual voice' have space and time for expression.

Done well, it will add a dramatic dimension to the other brilliant aspects of your presentations.

CHAPTER 10

Giving a brilliant performance

How you put yourself across to your audience is the key. No, you are not expected to be a professional actor but you are expected to be a brilliant presenter. You are expected to put on a show because that's what presentations need – a bit of underlining, a bit of *drama*, a bit of swagger and total poise.

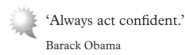'Always act confident.'

Barack Obama

But acting can help

Dustin Hoffman takes his job very seriously. When he was in *Marathon Man* he had to appear exhausted in one scene. Ever the perfectionist, he stayed up all night and arrived on the set the next day white, puffy eyed and knackered. Laurence Olivier surveyed him in amusement: 'Why don't you try acting, dear boy? It's so much easier.'

At an event in Bombay in 2009 at a ceremony called 'The Giants of Mumbai' (Bombay/Mumbai ... take your pick) philanthropists and high achievers were honoured for their work for the city in front of the local great and good. There was a lawyer, educationalist, businessman, academic, broadcaster, doctor and two actors from Bollywood, one of whom had arrived to female screams and lots of paparazzi camera flashes and clicks. Shahid

Kapoor is an actor: 30 years old, very good looking and knows it; a rising star. He spoke and here's my recollection of what he diffidently said:

'I have acquired a certain transient fame ... After my last film who even knows about that? ... But that fame allows me to open doors ... I want to do that for you ... for any of you doing the great work for the poor of Mumbai that you do. Thank you for that and for this award. I am here for you.'

What film actors have is the knack of speaking well, clearly and straight. Learn from them to be confident, poised and a good speaker ... it's a great start. As for Shahid, we were all blown away.

 **tip**

First of all learn to speak well – clearly and simply. And learn to respect your audience (as well as love them).

In praise of hard work (in praise of America)

While Britain was creating the aristocratic and amoral character Raffles, the Americans were creating Rockefellers. While Britain was defending an empire, they were building an economy. They worked incredibly hard and they still do. They don't dream the 'American dream' – they create it brick by brick. Their success is founded on an astonishing attention to detail. Whenever you hear people knock the USA, as many do, remember their work ethic and look at some of their achievements.

Which leads me to Al Pacino who, when he performed in and directed Shakespeare's *Richard III* in Britain, gathered the cast around and dissected the text with them word by word, examining meaning and motive until it became clearer and clearer what was going on. Shakespeare, he told his exhausted fellow thespians, wasn't casual or sloppy. If a word was there, it was there for a very good reason. Find it, he said, find it.

So it is with presenting – if a word is there it must be there for a reason. Getting it right will be achieved only by rigorous questioning and a refusal to stop until you get to the bottom of an issue. I recommend such rigour to you in preparing a presentation. If you work in an American company you will find that exhaustive examination of every word and nuance is commonplace. Although, and there is a downside, you'll also find a rather tiresome didacticism in their presentations. They tend to shout rather than converse.

 **tip**

Work as hard as you can on creating a solid, logical argument for your presentation.

Performance starts with how you sound

Your voice is crucial and while actors start with an advantage you need to improve and work on your voice, how it carries, how much variation there is in it, how resonant it is, how good your diction is. This is potentially your biggest asset – or liability. For most audiences it's all that exists of you when you give a 'whiz bang' slide presentation. The rest being hidden in darkness.

The key when it comes to training your voice is not to be shy. Learn some Shakespeare and declaim. Or, it could be Keats, Shelley, T.S. Eliot or Carol Ann Duffy – anyone you like. Just learn it and then try saying it out loud – in the bathroom to start with, then outside. Raise your voice and let go with exaggerated theatricality. Enjoy yourself. All I want is for you to be in total control of the content so that you can focus on delivery. For example, imagine reciting 'Ode to a Nightingale' to a child, to your parents, to an old lady, to a lover, in a large church to 200 people, in a sitting room to 12 people. The exercise is in learning how to deliver the same thing in different contexts and exploring your own range of expression.

So how do you feel about your voice? Personally, I like and I lean on mine. It sounds quite strong and it gives me confidence – I can play with its range and pace. When it isn't quite there, through a cold or a surprise attack of nerves, my sense of self-belief begins to flutter.

brilliant tip

Lean on your voice. It's your friend. Look after it with water and mints. Listen to it. Practise. After all that's what singers do.

Your voice is as important to you as a presenter as your putting stroke is to you as a golfer – so look after it. Record yourself and listen to yourself. Work with a coach if necessary – and it probably is. Someone who slows you down, who makes you speak in lower bass tones. Look at what Sir Gordon Reece managed to do for Margaret Thatcher – she was transformed from a Finchley housewife-politician who, as education minister, stopped school milk ('Margaret Thatcher – school milk snatcher') to a world statesperson who spoke with gravitas and passion. The late Edward Heath had voice coaching with the same result, producing in his case a statesmanlike burr. Current politicians try anything from nasal surgery to coaching to get their voice sounding more impressive.

Illogical arguments show up in a presentation

The question is 'Does the argument hold water?' *not* 'Is this good rhetoric?' Don't be sloppy in building an argument and in creating the logic of your story. At least one acid test of building a credible presentation lies in the practice of constantly interrogating the logic and the flow of the argument.

Don't go to bed until you've cracked it. Toyota created 'the five whys' as their process of inquisition whereby any proposal was intellectually tortured until it confessed to its weakness, or survived through the strength of its truth, convictions and logic. For example, suppose that a young executive plans to present a proposal to launch a range of premium organic chilled meals at the next senior management meeting. His boss is sceptical about the plan and the way his executive is going about it.

- **Why** are you making this presentation when our resources are already overstretched?

  Because research shows there's a big gap in the market for this concept, and it is very high margin so there's an incremental volume and profit opportunity. And I'll put in extra time myself to make sure the presentation is brilliant. The resource issue is down to me and my time.

- **Why** do such a high-profile presentation? There'll be sceptics there on our team who'll suggest the absence of competitive activity may suggest that our competitors know something that we don't?

  Because it's very hard to get the product right but I know we can do this if I can enthuse enough people on the team. Hence my wanting to put on a bit of a show. No one else has our organic credentials so competitors are wisely steering clear. Anyway, I think we're a bit smarter than them and I want to show that in my all-singing-and-dancing presentation.

- **Why** do you think that you can overcome your colleagues' antipathy, which you concede is a problem? They'll probably say if Tesco were to agree with your market diagnosis they could simply do it themselves – and probably would.

Because I want to be high profile on this and gain first-mover-advantage. You see, if we can get Tesco to stock, in addition to the easier targets like Sainsbury and Waitrose – and I think we can because the taste tests on our product are exceptional – then the organic story plus our pricing makes this a must-stock range.

- **Why** should we do this and run the risk of cannibalising our relatively successful, but static and under margin-pressure, non-organic offering?

  Because it's all down to positioning, isn't it? If we launch head to head against ourselves then you have a strong point, but if we aim more upmarket and go for people organically predisposed and avoid recipe duplication, then we should be OK. Especially if our marketing is as exciting as I plan and aim to show.

- **Why** would you want to put your career on the line now, just when you are on track for promotion? You see, if you go high profile, make this the presentation of a lifetime and it fails to hit plan, your life here is over, and I doubt if any of our competitors would be impressed either. Why not dip your toe in the water? No big launch. No glitzy internal presentation. Will you think about that?

  That's a very strong point boss and very well made. Can I go away and have a think about it? What sort of promotion did you actually have in mind by the way?

Why? Why?? Why??? Why???? Why?????

This technique will usually find the weak point in any argument, position or presentation. Try it on the argument in your next presentation. When, through laziness or because we can't quite make the argument compute, most of us make jumps in our logic, bury issues that don't suit the argument or, if we are politicians, we sometimes fib. Sometimes, we even tell really big porkies just to get a brilliant presentation away – sometimes, we believe $2 + 2 = 5$ in reality, not as a metaphor, because we just can't be bothered to do the simple arithmetic.

 **brilliant** tip

Sometimes we're tempted to tell really big porkies just to get our presentation flying ... or, if we're honest, we know the story we're telling doesn't quite stand up and we know it shows.

A strong argument helps you feel confident

Your performance deserves a strong, bulletproof argument (as opposed to just a bullet-point argument). Not only will it always be better if it has this, it will also be easier to perform – no more butterflies about being destroyed by questions in the Q&A because of corrupt logic.

So let's consider the ways in which your performance can be enhanced so that you stand out as a presenter. At this stage there will be those among you who say that you hate acting. You might as well say you hate work. Crudely, if you are an executive, learn to present by putting on a good show because all corporate life nowadays involves a degree of acting – as in maintaining a poker face, as in controlling your temper, as in being nice to people you don't really like, as in being patient with the personal woes of an employee, as, in short, doing your job. And any way you can find that helps the performance is worth taking. Even bribing yourself to do a great show.

brilliant tip

Reward yourself every time you do a major presentation.

Performance – left and right brain

Left brain is the functional stuff that obviously matters. The techniques that make you clear, audible and authoritative. Right brain is the creative stuff that, when it works, makes your presentation fly.

To start with you need to be sure you have the ability to achieve those functional competence issues and pass the left-brain test. You need to tick these boxes. Do you:

- look good;
- seem confident;
- know your stuff;
- seem very well rehearsed;
- come across as audible and clear;
- finish on schedule (and never over-run)?

And then the right-brain stuff that fills Malcolm Gladwell's book *Blink*. This is about first impressions and qualitative responses from the audience's viewpoint:

- Do I warm to this person?
- Do I trust this person?
- Would I like to meet this person again and carry on the conversation?
- Are they inspiring?

I've already talked about sitting on the metaphorical flight deck doing your pre-presentation check. Now, to achieve 'flight clearance' to do a presentation by your left-brain control tower, read the book, follow the tips, practise, work with peers to make sure they can see you, hear you and understand you.

For right-brain flight clearance you need to read and put into practice what's in this book. But you should also, when you've done that, work with a presentation coach who'll help you do the three things that make the real difference to your on-stage presence. They'll do the following:

1 Help you find your voice – the style, tone and pitch which seems to work best for you.

2 Help you be yourself but a bigger more dramatic version of you.

3 Help you simplify and strengthen the impact of your body language.

And you'll find the more work you do the easier this process of self-discovery is. There are some presenters who just love doing it and whose appetite for it, like a chef who loves food and the taste of it, is totally beguiling. Like Sophie Patrikios, Head of Consumer Services at Lego. Like the service Lego claim to provide, she's fun, knowledgeable and engaging. If you get the chance to see her do so. She's not just nerveless about it, so much as in love with presenting and telling her story.

brilliant tip

Find your voice. Find the 'you' that comes across best in a presentation. The more you do presentations the easier this is.

brilliant checklist

● **Be yourself:** David Abbott, one of the legendary advertising figures of the twentieth century, spoke of people going into 'presentation-speak, hunched up, glazed eyes, Dalek-voiced and didactic'. There's no need for that. Take what you have and make a vow to lift yourself up two or three levels. Be bigger. Impose. But don't be someone else.

● **Work on how you look:** Of course you need help, how can you work this out yourself? Get advice on how to dress – for women, a well-cut ▶

trouser suit always looks powerful and modern. (But in the end women have a much better sense of style than I have. My only comment is make sure you feel great and if that means buying a new outfit for every big presentation you make, then do it. It's a good incentive to be brilliant.) For men, a dark suit with a white or light blue shirt and no tie (though you can wear one if you want). Slight sun tan – no workaholic you. **Do not let anyone, however well intentioned, make you wear anything or have a look that makes you feel awkward.** The whole point about this is to make you feel better about yourself – you are the star. You need to think and feel that you look your best to be your best.

- **Do you have stage presence?** Look around you as you go on stage – the bigger the stage, the bigger the look and the bigger the sweep of your gaze. Walk on with a straight back. However you really feel, look as though you are glad to be there. It's worth repeating that – look as though you are glad to be there. Smile and say 'Ha!' under your breath (especially under your breath if you are miked-up) – it will animate your face. If it's a smaller event, simply try to be engaged and interested in what is going on and when you start or stand up to speak make eye contact with a few people and think 'Hey, I really like (even love) these people.' It's amazing how this communicates to an audience. The number of times I've seen ill-at-ease, shifty, grumpy or even downright hostile speakers is legion. Try to imagine that the people out there have loads of money and that they are going to give some to you, but only if you look cheerful and really glad to be with them. Stage presence is like sex appeal – some people have it naturally, some acquire it. Whenever Marilyn Monroe stepped into a room everyone stopped and noticed.

- **Try to radiate self-belief:** As someone once said to me, 'You have to believe what you are saying even if you don't believe it.' What is it that marks out the brilliant presenter? The best presenters in the advertising business tend to be shameless showmen. 'Shameless' is such a great word – say it to yourself next time you walk up on stage, 'Hey I feel totally shameless today', and see what happens. All you have to do is radiate confidence and a conviction about your subject. Lord Denning, one of the more formidable orators of the twentieth century and a fine lawyer, recalled his pupil master saying to him: 'People pay us for our certainty, not for our doubts.' Always act confident.

- **Never talk in jargon:** Something about standing up in public and talking has always (once I conquered my nerves) made me giggle. And in truth many presentations are absurd, especially in a world where jargon is rife. A few years ago someone invented 'bullshit bingo' – a game in which you drew a box within which were 36 squares. Each square contained a piece of management jargon – 'going forward', 'strategic', 'focus', 'blue sky', 'global', 'scale-up', 'benchmarking', 'culture', 'ratio', 'downsizing', 'outsourcing' and so on. The game was to mark off each loathsome word as it occurred and if a vertical or horizontal line was filled then the lucky winner leapt to their feet shouting 'bullshit'. Avoid all jargon, speak English – in short sentences with short words. And remember that the way a script is written bears no relation to normal writing – it should have short sentences, big type; it should be fast … conversational … bold.

- **Enjoy being a bit dramatic:** You know where you are, why and to whom you are talking. You have a good, simple, strong story and the logic is sound and bombproof. The slides are coming along nicely. Yet you need to impart some surprise or 'oomph' into it at, say, two points so that you stand out from your colleagues (in other words your competitors) and so that the audience remembers your message. Vivid recollections of mine focus on disasters: in the old glass slide days when the slide projector caught fire and the slides melted in front of us so that 'why we shall succeed' dripped into the reproachful word 'succ', which slowly blackened and disappeared; the slides that had been fed in back to front, making it look as though the presentation was in an eastern European language. **Think of what you most want your audience to remember and then brainstorm a solution that dramatises the point**. It could be a huge word on the screen, a piece of animation or video or soundtrack, a great quote – or you could remove your jacket, tie and shoes, saying 'This is about removing excess cost and baggage.' Whatever. It's about finding just one moment when you go for a bit of dramatic emphasis.

- **Don't be too mechanical:** Breathe life into your stuff. On the road to brilliance you have to learn to make what you say attention-grabbing and interesting. When you stand up, you have to look as though you care. You have to breathe life into the case you are making. You have to be passionate for your cause. You have to be (and be seen to be) glad to be doing your presentation. I was recently at an event celebrating an AFRUCA anniversary (Africans Unite Against Child Abuse). While there, I reflected how much better the Africans were at demonstrating passion, sincerity and involvement. The reason, I suspect, that so few leaders currently come from the UK is that we are hopeless at being extrovert – unlike the South Africans, Zimbabweans, Australians, Americans, French, Italians. So when you get up there give it some 'welly'. **Be animated – look and sound** *dynamic*.

- **Practise, practise, practise:** You will need technical run-throughs and you will need to know your material very well. Also, the situation may be changing, which means you may have to be nimble-footed enough to change parts of your presentation at the last moment. But most of all you must leave yourself time to explore performance – to work on pace, on pauses, on the louder and on the softer bits. Don't imagine that any good performance you've ever seen wasn't very well practised. They say Steve Jobs took weeks working on some presentations. Practise alone so you get used to your own voice. Don't do what some have attempted – editing and moving material around while actually performing. This is nerve-racking for technicians and rotten for the performance – what you have when you start is what you go with. Do at least two trial runs: the first is what I'd call a 'walk-through', just to make sure all the elements sit together; then a 'run-through' which is a rehearsal with feeling. Mark up your script with cryptic notes and dream the performance in slow motion. Have a plan B – there could be a power cut or a fire, maybe hardly anyone turns up, maybe you are the only speaker standing because of dodgy pasta at dinner that you alone missed because you were rehearsing. **There is no substitute for maximising your chance of success through practice.**

brilliant tip

If you're on stage we have no place for being shy. Aim for a power performance with loads of feeling, thoughtfulness and confidence. Everyone can be really good. It just needs hard work and a lot of thought. Nearly all brilliant presenters are made, not born.

brilliant recap

A bit of theatricality is vital if you want to be brilliant. Even if your performance is controlled you are still on stage and you have to fill that stage. You are not there to talk or to read a script – you are there to command attention, engage people and convince them about your story. Tom Peters and Jim Collins have been engaged in an energetic debate about styles of leadership. The former advocates executives to be charismatic, noisy, short-term and disruptive; the latter is all for quiet, reserved, process-driven, data-rich, team-building and long-term.

Each has a point. We probably need both but, arguably we don't have a 'long term' any more – we have *now*. And if you're on stage we have no place for the 'quiet man' – remember Iain Duncan Smith, one-time leader of the Conservative Party in the UK? Aim for a powerful performance with lots of feeling and loads of confidence. But, back to Iain Duncan Smith for a second. I recently saw him perform at a small gathering talking about poverty in the UK and the role of welfare. He was very funny … proving his credentials as a human being up front and then without notes, he gave a brilliant exposition. He really knew his stuff and spoke persuasively and with supreme confidence.

The 'quiet man' had transformed himself into a brilliant presenter – yes, it can be done.

PART 4

How technology and technique impacts on presentations

W e are so lucky. Suddenly we are confronted not only with magical technology that makes animation and video easy to embed in a presentation, but we are also exposed to some very talented, brave and exciting presenters who are pushing back the barriers of communication with new and interesting techniques. Technology and technical proficiency are aligned in symphony.

A word on PowerPoint

Since the invention of PowerPoint 24 years ago a number of things have changed. It has been adopted by all businesses throughout the world, however grudgingly by some. It is commonly used in schools and universities. It has been adopted by churches, community groups and governments. Everyone has the technology at their fingertips and can and does produce PowerPoint presentations. In fact the two words, PowerPoint and presentations, have become synonymous. Many people judge themselves by their special effects. PowerPoint is as global a business tool now as double-entry bookkeeping.

When technology hits the stage

This part explains how to improve the technical side of presenting underlined with one big caveat: a bad story, badly put together and badly told will never be saved by clever technology.

Having said that, if you want to put on an impressive show then you are in charge of an armoury of sight and sound none of your predecessors commanded.

Technology and technique

The part falls into three sections:

1 The basics and developments of PowerPoint, Keynote and other technology.

2 Staging, lighting, sound, taking the show on the road and insights from behind the stage.

3 The techniques used in live-presentation and on TED by some of today's most accomplished performers.

How to get the most out of this

Never be held in thrall by PowerPoint. Yes, it is a useful tool and can be a good friend, but it can also be a complete pain (mostly for your audience) if you are dominated by it. Technology can make you sparkle or – if misused – can simply make you look flash.

The most useful thing you can do is to study as many presentations as you can. Become a presentation junky. Study what presenters do on stage and how what they do seems to affect the audience. **Look, listen and learn.**

CHAPTER 11

Technology
is here to
make us more
powerful

PowerPoint is regarded as a critical and brilliant tool by many presenters. Yet opinion about it is polarised – I regard it as a brilliantly useful tool that can focus thinking and enable you to produce, at least, easy-to-assimilate handouts and, at best, swinging slides. Even so, it's just a tool. This chapter is more about process than anything else. But do remember that if you want the best from your career-shaping presentation then create a relationship with a professional slide designer and marvel at the difference you get between, as it were, your fast food and their cordon bleu cooking.

PowerPoint, friend or enemy?

Because of its ubiquity, PowerPoint has become the generalised term for all slides. In fact it means:

1 the discipline of using rectangular computer-generated displays on a screen or in a handout to enhance the presentation (using PowerPoint, Keynote and other specialist presentation programs);

2 the particular Microsoft program with its built-in formatting styles – most notably bulleted lists of points. But note that Microsoft themselves have published a book called *Beyond Bullet Points*.

There are, we're told 30 million PowerPoint presentations done every day – or, if you prefer, 350 PowerPoint presentations start every second. That's 1000 since I started typing this sentence (I'm a slow typist). Around 11 billion presentations a year are hitting the population – that means there are about twice as many PowerPoint presentations a year as there are people on the planet. There are PowerPoint presentations everywhere.

It started in Silicon Valley – a man called Bob Gaskins with a doctorate from Berkeley had the idea. He developed it with Dennis Austen at a company called Forethought. PowerPoint 1 was designed for Apple Macs in black and white, followed quickly by a colour version. They were bought up by Microsoft at the end of a manic year for $14 million. Bob must have thought 1987 was full of Christmases. I wonder what he'd think now.

Pluses and minuses

The existing software has its advantages:

- It makes people begin to think visually.
- Used at its simplest it can be very effective.
- It's great for consistency, leading to a coherent, corporate feel.
- It can create greater memorability of the presentation as a whole.

However, it has weaknesses too (mostly because it's so easy to use):

- It encourages the use of too many words.
- It is very easy to manipulate, which generally leads to too many charts.
- Too many people are doing poor experiments with it (poor for the audience, that is).
- Too much time is wasted on tweaking.
- PowerPoint presentations are generally ugly because people are not trained in design.

Blame the workman, not his tools

The apparent weaknesses of PowerPoint are about its misuse rather than anything being intrinsically wrong with the tool itself. It's like blaming the ballpoint pen for producing boring documents. You really can't blame PowerPoint for the bad judgement of those who jump in and misuse it.

PowerPoint is easy to use, but very hard to use well. It's like dancing – any fool can jive, but to see the full potential, watch a professional do it.

PowerPoint is like 'painting by numbers' – it gives the illusion of craftsmanship but in fact it just gets you on the road to painting. It allows anyone and everyone to have a go.

PowerPoint has improved working lives

In the so-called 'good old days' we used slides of artwork that cost £20 each and took experts two days to prepare. Or we had overhead transparencies (which are still a fine tool for small groups – you can even write on them).

When we moved to computer-generated graphics, the first projectors took hours to set up (three guns, each of which needed to be put in focus) and were very sensitive – one knock and you were done for. They cost £25,000, about the same as a concert grand piano. Now a superior digital product with a single lens as opposed to three, with auto-focus, costs about £500, the price of a reasonable electric guitar.

Click on PowerPoint and there they are – a set of 30 templates. There are only 3 or 4 that I find useful – you type in your presentation content and in about half an hour you have a half decent draft presentation. However, this dish is still raw.

You then put your slides into notes page format and work on simplifying the slide, above what on the notes pages is beginning to look just like your script. As with sauces, you reduce and you reduce. You are an intelligent human being and this is just the beginning of a long, iterative process.

Those who don't like it describe such presentations as 'death by PowerPoint'. However, from a presenter's perspective it's 'life by PowerPoint' because the relatively PC-illiterate among us can still work, refine and improve all through the night without backup. We have been liberated.

brilliant tip

Watch television to see how experts develop graphics and whether you can borrow their ideas.

The way the world is moving on (and sometimes not)

When you talk to an expert they make it all sound so easy. I was interested to hear how the MD of Showcase, Martin Conradi, saw things from a technical standpoint, having spent so long in the forefront of changes in the presentation world.

Like Steve Jobs he really believes that you have to start with the audience experience and work backwards from there. If the audience can see how clever the technology is then you've probably got it wrong.

All programs can link text, graphics and video together and output them in a variety of formats – so there is little to choose between them from an audience impression/satisfaction point of view. In the hands of professionals the results are virtually indistinguishable so conversations about which option to use aren't that helpful.

More interesting are display options. The standard 4:3 screen is moving to 16:9 or even 32:9 for conferences; for a long time it has been possible to configure video walls and LED blocks (think in terms of Lego) into all manner of irregular shapes. LED blocks can work effectively in high levels of light provided the screen size is large enough to mask their low individual resolution, but resolutions are improving all the time. Holography has been mooted but is still years away. However, screens made of fog allow you to walk or drive through them and clever optical tricks with glass can appear to audiences to show humans interacting 'live' with computer animation.

Video and beyond

Simple video-style effects are achievable but for anything complex in the near future they are expensive, slow to produce and beyond the scope of most presentations.

iPads and other tablets allow a presentation to be controlled by a speaker on the move, although compatibility is an issue (for example, only limited Apple Keynote functionality is available on an Apple iPad at the time of writing). They also allow the presentation to be wirelessly distributed to individual iPads which in turn can be connected to large screens or projectors, with the speaker controlling the slides from their own iPad.

Other presentation software is available, often claiming to be 'PowerPoint compatible' – though this often refers only to the most basic PowerPoint functions. Importing or exporting real-life presentations into or out of PowerPoint can be disappointing and frustrating. So beware.

Presentations can be enhanced with images taken from the web (go for high resolution and beware of copyright) or photo libraries. Or you can enhance it with artwork created with graphics software such as Adobe's Photoshop, Illustrator or After Effects. Many programs designed to create web animations such as Flash, Silverlight or HTML5 can be used with presentations but may need special drivers. Alternatively the presentations themselves can be converted into Flash and other formats which can be helpful where the presentation playback software may not be available.

Such is the flexibility of the latest versions of presentation programs that a PowerPoint 2010 presentation can be saved as a WMV video and then embedded in itself (this is rather impressive but not necessarily useful).

You want to be flash? Listen to Uncle Steve

Apple is intent on excluding Flash from running on its iPads and iPhones, expecting HTML5 to increasingly take over (check out Google on 'Apple thoughts on Flash').

You can create entire presentations with many of these programs – Photoshop or Flash, for example – and they can be very impressive. But they are time-consuming to make, and slow and difficult to edit. Great for presentations with long-term, multiple use, but not for anything likely to need changes at the last minute.

Any video or animation on screen takes the audience's attention away from the speaker. Fine if that is what you want but it should be used with caution.

Showcase Production secrets

Create a presentation world where everything is big and nothing is small.

1 Get the story right. What is the one big point you want to get across?

2 Cut down the words to the essence. Keep it simple. Less is enough.

3 Find pictures to illustrate. One big picture per slide, not lots of little ones.

4 Find a visual style. (A Dorling Kindersley executive once did a presentation about the challenges of getting good images of elephants on a white background. It was an elephantine and never-to-be-forgotten tour de force. One set of images defined the visual presentation and it worked.)

5 Keep it consistent ... except where, for effect, you want a discordant moment.

6 Beware language. In a Tesco Poland conference they thought they'd have a slide on pricing in the music industry. Their slide 'record prices' startled a few people.

7 Beware lack of line breaks on slides. For example:

The French push

Bottles up Germans

has a very different meaning to:

French push bottles up Germans

8 The technician running your show has you by the balls for the time you are talking. He may have been up all night building the set and wiring up the audio or lighting or whatever. Treat him as an intelligent, caring supporter; explain what you need and if possible run through it with him. Thank him at the end. This cannot be said too many times.

brilliant tip

Is the main audience live or on the web or the handout readers? You must decide which to go for.

brilliant dos and don'ts

Do

✔ Assume that you are very busy and have to get this presentation cracked quickly.

✔ Set up your template by starting with the Slide Master so that your style – font, bullet points, indents and so on – is consistent. Make sure it looks how you want it to look to start with.

✔ Write a contents page – this gives you a format, and a series of signposts in the right order. This will vary from situation to situation but it helps the audience if they know that you actually know where you are going.

✔ Always have a heading on your slide that says 'Strategy', 'Objectives', 'Challenges' or whatever but never more than five or six words – never. And always in lower case – it's easier to read.

✔ Keep it all cryptic, avoid jargon and use simple visuals.

✔ Focus on refining and reducing your presentation – it is your musical accompaniment and it's there to make you look good and your meaning clear. It's there to help you.

Don't

✗ Get obsessed with the pictorial side of things or elaboration – focus on the story and the overview.

✗ Use more than five bullets or 20 to 30 words a slide (preferably 20), except for handouts.

✗ Forget to leave gaps for pictures and diagrams, which may come later.

✗ Worry: you will always overwrite to start with so long as you promise to edit ruthlessly.

✗ Rely on your own judgement: if you have time, pass it to an expert to tidy it all up, and at least take out the ugly bits – better still, ask for a professional touch.

Technical advice to make your DIY life easier

Here is a list of things you need to do to stop you making stupid mistakes when making your own slides:

- Do not use subheads and sub-bullets.

- Do not have complex charts – five simple slides rather than one very complex one is preferable.

- Do not muck around with fonts and colour schemes – choose a simple scheme and stick to it. Ariel and Times Roman are pretty well foolproof – fonts such as Tahoma may not be. If you insist on unusual fonts your lovely presentation may go awry with strange line breaks and with half the text lost.

- Avoid animations, transitions and sound effects unless you have expert backup.

- Do not type your script on screen and then read it out – have pity on the poor audience.

- Do not include elements that may not work on other computers – taking risks with technology is like accelerating on black ice, a recipe for ignominy.

- Do not assume that everyone has an up-to-date PC like you. Having said that, do buy an up-to-date laptop, which you will henceforth call an 'audiovisual backup device'. Do not ever call it a computer – IT departments get funny about stuff in their territory. AV devices are someone else's problem, especially backup devices, provided they don't impinge on the IT budget.

- Recognise that there is a difference between producing slides for projection in a big room and handouts for takeaway. The two are not the same, but many people assume they are.

Things to do to your PC (if you know what you're doing)

1 Modify your toolbar so that it is easier to work with – you don't need half the stuff there.

2 Use the Slide Master, which you'll find under Master View – it enables you to store information about font, positioning for text and objects, bullet styles, background design and colour schemes to be applied to all your slides, including any you may add later. It also allows you to apply retrospective global changes – such as changing font or colour – to all your slides.

3 Under Tools/Options/Save you'll find 'Allow fast saves' – turn it off because it inflates the size of the file, as well as making it more prone to corruption.

4 Use PowerPoint help groups (for example, **http://office. microsoft.com**) for tips and shortcuts.

5 Find useful shortcuts and keep a list of them – for example, Shift F3 turns everything into capital letters, Ctrl B or I or U formats a highlighted section as bold, italic or underlined respectively.

6 Check your presentation prints in grayscale.

Practise and be clear about what you want

Most important of all, if you want to be competent with PowerPoint, practise, practise, practise and never stop practising. Use the following websites to get advice, help and ideas:

● **www.showcase-online.co.uk**

● **www.rdpslides.com/pptfaq**

If you bring in professionals to help with your slides, staging or script, brief them properly. There is nothing more frustrating and expensive than expecting outside suppliers to read your mind. And, however good you are, never stop working on finding ways of getting better.

brilliant tips

- Practise PowerPoint. How can you expect to be any good unless you practise? Those with real talent start with two distinctive things, an eye for design and a job – which means they do little else other than slides morning, noon and night.

- Exercise self-discipline. I mentioned earlier the 'corporate porridge' imposed by corporate design templates. Well, unless you are the boss, stick to the rules and work within them. There are no prizes for being radical when radical is illegal.

- Work with experts. Learn techniques by working with PowerPoint experts. At very little cost, they will show you shortcuts and effects you haven't understood before.

- Study design. Look at the BBC for instance – its journalists and designers don't use PowerPoint but they are usually good at design. So is the *FT*. So is *The Sun* – study how they make their audience focus. We are, as a bunch of business people, hopeless at design. Become a 'design junkie' – it's one of the skills for which the UK has a strong reputation.

- Use notes pages. These help you to get that critical point of distinguishing between what you think *you* need to say and what the audience needs to see.

- Fire the bullets. Try to do a presentation free of bullet points and see what happens – you might liberate yourself from the 'bam-bam-bam' effect they tend to have.

▶

- Find out what you can't do. If there is any really legitimate criticism of PowerPoint it's that it masks incompetence, not just from the audience but also from the presenter. You need to discover what you can and can't do if you are going to get better at it.

- Focus on the 'big one' to start with. This is the Jack Welch thing, which says you need a killer strategic slide – one that clarifies your message in a devastatingly clear way. Learning to use PowerPoint can help you to break your thoughts into logic-sized lumps. Thinking more positively about visuals in presentations can take you from average to excellent, or apprentice to brilliant, really quickly.

- Never confuse handouts with screen works. Yet everybody does. We live in a laptop-sized world and when it comes to a larger audience we sometimes can't understand why what looked great in the office at 11.30pm on your PC looks dull in the Queen Elizabeth Hall the next day. Big is not beautiful – it's merely *bigger*. And the really embarrassing thing about bigger is that it magnifies amateurism. What we produce for handouts and what we show on screen are not the same.

- Never use the word 'deck'. This is the word made famous in the 1970s when IBM was more famous for its presentations than commercial success. Why does it worry me? It sounds big and heavy – as though you could stand on it. It also sounds violent – 'I'll deck you.' What it doesn't sound like is fun.

You have only one chance

There is only one 'first night' when you do a presentation. You can't come back the next day having ironed out the wrinkles. This is your Terminal Five moment. You can change anything *now* – but you can change nothing afterwards (apart perhaps from your job). So practise now, as though your life depended on it. PowerPoint can liberate and enable you to visualise how to make a presentation work brilliantly.

But don't be an idiot by being a DIY bodger. You can do a small-scale presentation off your own bat but, however much you play with it, don't try doing a bigger performance without design professionals. If you do, you'll regret it. After all, you are the script writer and performer. Get the professionals to do set design, effects and stage management – under your direction, of course.

You thought technical stuff was easy?

So you have a plan, you have a story and you even have some pretty good slides. What can go wrong?

- The venue is dreadful with pillars that get in the way. You did check the room yourself? Ah, you left that to an assistant. Hmmm!

- It's got terrible acoustics – erriblecoustic … oost … ic … You did at least check this, didn't you? You didn't! That was like buying a suit without checking the size and without trying it on first. Lunatic!

- The projectors are underpowered so your subtly coloured slides look washed out – a bit like you'll do by the time you've finished your show.

- You decided to present using a laptop screen to 20 people. Huge error – tiny screen and an irritated audience.

- Headline fonts should be 44 point or thereabouts and text should not go below 20 point. Hard-to-read charts are the cardinal sin. Bravely, you ignore this and watch in dismay as all these myopic 60 year olds walk into the room. Pity but they won't be able to read the slides at all. Are you in trouble!

Anyway, you're ready to roll – here are some reminders.

 recap

- Don't read the slides.
- Have cryptic notes on your notes page, which you use as your script guide.
- Writing good scripts is a *very, very* different art to writing an article, an email or a book.
- Your script is there for you and no one else – so have it in huge print and short sentences.
- Avoid big slabs of text – you'll lose your place.
- The best presenters go through a five-stage process:
  - think;
  - notes;
  - script;
  - cryptic script;
  - notes.
- Always write your own scripts because reading other people's scripts is lazy and will consign you forever to being a boring novice.
- Do not allow yourself to over-run – this is a crime equivalent to burglary because you are stealing the audience's time.
- Pauses are good … Long pauses can have a dramatic effect. Important though the audience is, make it clear that you are in charge of this pause and that you haven't simply dried up in blind terror. One speaker employed the dramatic pause too liberally and someone in the audience stood up and asked 'Are you all right?'

But most important of all are these three things:

- **Plan** so that you really know what you are trying to do.
- **Rehearse** so you leave nothing to chance. This does not mean going through your presentation so many times that you get bored with it. It means knowing it well enough to know what comes next and to know where the high spots come.
- Spend twice as long as you are used to on preparing presentations and do as many as you can. It is only with practice that you'll improve. Have you, for instance, ever heard of an accomplished linguist who spoke French, or whatever, only once a year? The harsh truth in life is that you have to work hard. As the golfer Gary Player said: 'You know, the more I practise the luckier I get.'

However, it also makes great sense to check regularly what you are doing right and not-so-right as a presenter. Do it with a professional, someone who is constantly exposed to different presenters. This will be refreshing and expose the faults that in isolation may not be a big problem but which in combination may be holding you back.

The final, critical point – enjoy yourself! No, that's not a joke. But how, you ask, how can you enjoy standing up there with shaking legs and the strong feeling that you now know what the onset of a heart attack feels like? By getting used to being good – that's how. And when you've done a brilliant presentation, just the once, the euphoria will make you feel amazing and you'll find that you won't be able to wait for your next outing.

CHAPTER 12

Technique: learning from the best – the X Factor

L earning from the best around and understanding what they are getting right (and wrong), what their strengths are and the ways in which they achieve their impact on the audience is the quickest way of moving your own performance along. A senior executive with whom I'd been working went to a global conference in Hawaii and described seeing presenter after presenter employing the tricks described in this book. She said she'd learned especially about the focused use of passion (not a splurge of enthusiasm so much as a laser ray of energy) and the presence of memorable splashes or strokes of colour.

The presentation game has been developed by some significant performers in the past few years, most notably Barack Obama and Steve Jobs. The online (and life-performance) phenomenon TED and the explosion of serious speaking and debating events sponsored by organisations like *The Spectator, The Week,* the Royal Society of Arts, the Royal Geographical Society and others, have also made an impact. Increasingly the work of academic institutions like the London Business School, the London School of Economics and European institutions, like IMD and the University of Lausanne, in promoting their institutions and parading their star speakers are changing presentations into new stand-up entertainment.

The power of the spoken word for a generation who are told that the golden days of parliamentarians is over, is dramatically demonstrated at an increasing number of very well attended events. The fact is, the art of presentation and storytelling has never been in better or more experimentally vibrant health.

Yet in business presentation skills lag behind

MBAs courses do not include presentation or communication modules. While most executives willy-nilly find themselves doing more presentations than their predecessors did, the actual art of business presentation as opposed to speech making and debate still belongs some way back in the last century. When a particularly dull presenter said to me somewhat patronisingly that there would be no need for another book on so marginal a subject as presenting, I knew the world was in trouble and I should need to argue the case for presenting strongly.

 **tip**

By hard work, intelligence and energy you can still stand out as a business presenter.

Enough groundwork is now being laid with the young at most schools and through the presentation organisations I've described to suggest there's a fundamental shift in opinion and practice. It's when the world realises that presentational skill may swing a shareholders' meeting or a tender that the real swing in opinion will confirm presentation coaching as a mainstream activity. The skill of a Stuart Rose, or more recently Marc Bolland, are no longer seen as 'one-offs'. The increasingly adversarial confrontations at investors' and shareholders' meetings that major institutions face put the senior executives' debating skills top of the executive development agenda.

Achieving that *X Factor* performance

It's time to mark a few cards and to reflect on the performances I've seen over the past 30 months or so at the hundreds of presentations I've sat through. I'm doing this on three dimensions:

1 The overall impact of the event in terms of organisation, venue and theatrical impact.

2 The best of the performers and why and what we can learn from them.

3 The individual styles and why one size will never and never should fit all.

The overarching comment is that there is a consistently higher standard of delivery than we saw 20 or 30 years ago. Expectations are higher on all sides. When something goes wrong – like a video that won't play – there's mild irritation but there's an inclination to move on. Everyone is a bit cooler. The event is less capital 'E' than it once was.

Perhaps the most helpful observation comes from master presenter and highly paid performer Sir Ken Robinson, star of TED, guru, author and raconteur on education. He is good enough and smart enough to take the critical opening 10 minutes or so of his talk establishing a rapport with his audience and thinking about how to frame the rest of his hour-long or however long session with them so as to maximise its impact. This is the ultimate in having a speech in beta-form and researching audience need on the hoof. This really is X factor.

Places and ambience

I've recently seen presentations and performed in conferences as far flung as China, Lord's Cricket Ground, the Royal College of

Obstetricians and Gynaecologists (in Regents Park and a brilliant place, if a little grand) and a luxury hotel in Dublin under an Icelandic ash cloud.

The choice of venues nowadays is pretty well infinite. Whereas at one time Oxford and Cambridge colleges, the Parliament buildings, the Royal Opera House, Twickenham or Historic Royal Palaces were rather snooty about opening their doors to commerce, today business is business and, for a price, previously hallowed portals are flung open to all of us.

Three key lessons seem to apply:

1   There is seldom a perfect venue but most of them are very good. Provided the service levels are adequate and the place is well lit, clean and audience-friendly then the rest is down to the programme. If a venue has a touch of class, like Lord's or the Royal Opera House, so much the better.

2   It's the way in which the event is set up and how the expectations are managed that matters most.

3   The organisational details matter and the real coup is to score well for distinctiveness that is relevant. Thus:

 - Bloomberg's conference rooms for great, light, modern buffet-style food (breakfast and lunch);

 - the Royal College of Physicians for the most comfortable seats in London and the most impressively raked auditorium;

 - Lord's for the sense of space and the green grass through the windows;

 - Mercers' next to the Tower of London for that view (views matter: spend eight hours underground at a conference and the ability to think creatively dissipates pretty quickly);

 - Gleneagles Hotel for great space and fabulous break-out rooms.

At the end of the day the real key is not how good the sushi was or how unusual the place. (Sally Gunnell has her own small conference centre at her home in Steyning and she serves the coffee … so is a gold medallist asking you 'One lump or two?' impressive? If it isn't a great conference the sugar won't sweeten it enough.) What we all remember is what was said and how it was said. We expect things to go right and regard failures of scheduling, technology and so on as avoidably amateurish.

 **tip**

Have enough people to help distracted delegates. (TED are great at this as are the London Business School – very professional.)

The best of the performers and why

The big changes are towards:

- content and intelligence – thank you, TED, thank you, Barack;
- self-confidence and command of audience – thank you, stand-up comedians;
- radical business thinkers – thank you, Seth Godin, Martin Lindstrom, Malcolm Gladwell.

In the good-old, bad-old days a speaker stood up at a conference and delivered a long commercial for their organisation. I recall seeing speakers from Heinz and Burger King doing this and feeling miffed. And potentially entertaining speakers, erring in favour of discretion, changed from brilliantly indiscreet government advisors over lunch to stubborn mutes on stage whether Chatham House rules applied or not. These days still apply in China (perhaps unsurprisingly) where in a Q&A that I saw, the speaker, when asked an innocent question, blurted out 'No comment … I have nothing to say.'

Great experiences

It's intriguing to see the Olympiads at work. Most of all to see just what presentation coaching can do. They are all good. The worst would get a bronze in communication, the best gold medals in presentations. And this has all been learned.

Sally Gunnell

Gold medal winning hurdler – she shows a maturity of learning about herself and competition. Brilliant in telling the lead up to the big event. Even better on how she broke a world record when feeling like she was dying of flu.

Lesson: Tell it like it was and reflect on what was really going on. The power of honest self-evaluation.

Colin Jackson

Silver medal winning sprinter, Colin is poised, charming and easy-going without a shred of tension. He is an ambassador for intelligence, sport and achievement. He tells great stories about that village and what the once-in-four-years event really feels like.

Lesson: Act small, make your success the audience's to share and, as Obama said, 'Always act confident.'

Extraordinary discoveries

Del Roy

A research fellow at the Massachusetts Institute of Technology. His area of specialty is the neurological aspects of learning. To help he put cameras and mikes in every room of his apartment so over three years he and his team could study pretty well 24/7/365 the learning process of his baby son. (Thus we see how the word 'gaga' morphs into the word water.) But it's the child's first steps that are a presentational gold moment. As the child makes a tentative second step, he whispers 'Wow!' It's complete magic. Del uses his material with brilliant reticence. It's 'you-could-hear-a-pin-drop' material. On TED.

Lesson: If you have great material, be sure to keep it simple and short. Let your props (film in this case) do the work. Be humble if you are this clever.

Jill Taylor
A brain surgeon who described her stroke. It's a terrifying and riveting story. At times it's uncomfortably emotional but a supreme tale of analysis battling with catastrophe.

Lesson: Be personal. Be specific. Be graphic. Tell it like it is.

Insights into life
Jayne Haines
A senior HR person at GlaxoSmithKline. She loves work, her young family and running. She tells how she was warned off winning races at school – 'Let the others win too.' And she talks about how running every morning at 6am clears her brain and makes her creative. She has work/life/running sussed. Up at 5.30am. In work by 7.30am. Home by 4pm. Bed by 9pm. She describes the concept of running as her 'red thread'.

Lesson: Note-less. Prop-less. Stands centre stage quietly. Honest. She gives fascinating insights into a calm, confident and competent mind.

Matthew Taylor
CEO of the Royal Society of Arts. Before that he was head of the Number 10 Policy Unit under Tony Blair. He's a brilliant think-on-his-feet speaker. He talks about our having prehistoric brains in a modern world, about society working with the wrong kind of incentives and about how to get groups of people to work effectively together. This last is the burning issue of our times. He could speak for hours. His father Laurie Taylor is a famous broadcaster. Talking is in the genes.

Lesson: Be nice but if you've got it, flaunt it. By being as open and good natured as Matthew we all want to share in his thinking without being intimidated by it.

Ken Robinson

He's the master storyteller and the proponent of teaching the arts with pride. He chats with the audience, constantly asks them questions and digs fun at himself. Hear how cooking eggs for men is a singularly focused task.

Lesson: He loves talking with people about things and stories he values. The constant use of personal and other people's real stories hold what he says together. Anecdote beats fact ... well, doesn't it?

Passion about life

Melvyn Bragg

He is extraordinary. He's the British voice of the arts. From creating and hosting the *South Bank Show* on TV to serial writer of books about language, achievement and our island story. Scene: the Charleston Festival, East Sussex. A marquee, jam-packed with 400 perspiring devotees (it's very hot). He talks about his latest book *The Book of Books* about the King James Bible. An hour of passion. It's an hour on poetry, democracy and British history. He's like a jazz player going off on riffs about Richard Dawkins (the atheist) and the folly of the new Bible translations.

Lesson: If you have a passion feel free to let it all hang out. Learn how to talk with apparently unscripted enthusiasm. Speak from the heart.

Eric Whitacre

Composer. Classical conductor. Looks like a rock star and talks like a very intelligent film star. He's engaging and successful. He understands young people and music.

Lesson: If you look good and feel good, get up there and let it show. If you have a great cause, don't refrain from recruiting converts.

Sal Khan

He's gone from hedge fund analyst to super maths teacher. A story of why everyone should be good at maths. If they gave you 80 per cent at cycling that would mean you were proficient in everything except, say, braking. You need 100 per cent at each level of maths to be a player.

Lesson: A disarmingly simple story told with the passion of knowing he's right. I'm a convert.

Andy Stefanovitch

I've mentioned him a lot. He's wild, disruptive and has a love affair with change. He speaks at 100mph and regards inspiration as the only way to a better world. He's determined to drink it all in … every last drop of excitement, curious stuff and innovation.

Lesson: He's tiring but the effect is mesmerising. He's a role model for enthusiasts with a pile of excitement to describe.

Dynamic interpretation

Hans Rosling

The Swedish Professor of International Health at the Karolinska Institute. He's also a genius with data. He's produced a dynamic programme of various trends in poverty, health, population growth and so on which makes ice hockey look ponderous. He dances in front of his screen getting very excited about the speed at which China is catching up. Essentially, viewed like this statistics are beautiful. Check him out on TED.

Lesson: Being animated and enjoying what you are telling and showing your audience is enormously engaging. Producing knockout visuals can be a winner if you know how to behave with them. Hans is world class.

John Knell

Said to be one of the UK's leading thinkers on change and organisational development. He's founder of the Intelligence Agency and he's very good on stage. He talked at that big Arts Conference cogently saying (hurray) as a nation we need to look after ourselves better and (at last) that all this talk of collaboration is all very well but, actually, we are unpractised at it and not very good at it. A dynamic voice of intelligence in the midst of an intellectual mess.

Lesson: If you are intelligent, act intelligently. Deal with big issues. Say what you think. John Knell and Matthew Taylor come from the same stable.

Showing richness of thought

Carolyn Porco

Head of the Imaging Team on the Cassini Mission to Saturn. And she gets very excited by the unique pictures of Saturn's moons. It's impossible not to share her awe and excitement. A great story told with childlike enthusiasm.

Lesson: Be as excited as the story you have allows. If it's a first or a breakthrough make sure the audience shares in the 'wow!' impact too.

Chris Bones

Professor of Creativity and Leadership at Manchester Business School and Partner at Good Growth, he was Dean of Henley Business School when I first saw him present. He's a fluent walker around in front of simple, provocative slides. Is 'inspiring leadership' an oxymoron, he asks? He defines leadership as a contact sport requiring 'influence, judgement, drive and self-awareness'.

Lesson: Enjoy having a conversation with your slides. You are paid to be provocative as a speaker.

Margaret Heffernan

CEO, entrepreneur and author who talks winningly about women and the qualities they have at work. She's calm, logical and authoritative.

Lesson: If you have a good story about an ostensibly emotive subject, tell that story dispassionately to get your audience listening, and sprinkle it with lots of real life and personal anecdotes.

A helicopter view of the world

Bobby Rao

Ex-Marketing Director of Vodafone and now a partner at Hermes Venture Capital. At London Business School he gave a stunning overview of the collision of three markets – internet, consumer electrical appliances and media. He's a cool, fluent presenter who sees the big picture.

Lesson: Try to get above the detail and give your audience a perspective. Give them your experience.

Mike Geoghegan

Ex CEO of HSBC. Described as the Geoffrey Boycott of banking, he gave a world overview at the RSA which focused on the next generation growth economies – Chile, Indonesia, Vietnam, Egypt, Turkey, South Africa (also known as the CIVETS states).

Lesson: A grand world tour from someone who knows it is fascinating. I loved his describing Brazil as 'basket case to bread basket in a decade'. Sound-bites work.

Phil Redmond

A leading TV writer and producer known best for *Brookside*. He was also the inspiration for Liverpool as City of Culture. At the big Arts Conference he spoke with Scouse irony and truth: 'There's a lot of knee jerking and a lot of talk about the arts.' He is a businessman and it shows ... like his radical idea of putting all the good pictures we have on display instead of hiding them. A wake-up call.

Lesson: Here was a great example of Redmond playing a bigger Redmond. He could and did say what he thought. Truth is so powerful.

A wonderful person

Amy Williams

She won our only medal at the Vancouver 2010 Winter Olympics. A gold on the terrifying 'skeleton'. She's a beautiful and brave 29 year old. And she speaks of the experience inspiringly (but nervously because she's new to presenting) focusing mainly on the team behind her. She's very much a 'we' person. She uses notes and is a brilliant storyteller with (let's face it) a brilliant story.

Lesson: Tell the story straight if you aren't a professional and let your diffidence give it colour. Her own sense of surprise at what she's done couldn't be made up. Be yourself.

Sean Meehan

Professor of Marketing at the International Institute for Management Development in Lausanne. He spoke at the Visa HQ in London about his book *Simply Better* and about the world of marketing. Anyone can aspire to be an Apple, he mused, if they relentlessly improve. He's a charmingly confident speaker who's impossible not to enjoy.

Lesson: I asked about finding his voice. He said it derived from a lifetime of public speaking. Now every performance is evaluated at IMD and changes are made to get it better. Simple slides. No notes. Be respectful of host and audience. Presenting doesn't change but occasions for it do.

Very funny speakers

Sir Ranulph Fiennes

He is a one-off. Self-deprecating. Ironic. British old school. Possibly mad as he described losing alarming amounts of weight on his trans Arctic and Antarctic trek and 10 fingers and toes. Question: 'Surely there's an easier way to earn money?' Answer: 'Name one.' He's wonderful.

Lesson: Be a bigger you. Create surprises with your slides. Tease the audience. Underplay bravery.

Gavin Stride

Chief Executive of Farnham Maltings, the creative centre for the performing arts, Gavin is an iconoclastic figure in the Arts. At the Arts Conference he sat grumpily (yes, you can present sitting down) and said 'The big Society, or Society as my Mum used to call it…' and made dagger-sharp intervention after intervention. He's a very funny man through observation not through jokes.

Lesson: Sharpen up your sense of observation and your sense of curiosity like Gavin.

Speaking to me

Khaled Tawfik

An MBA student at the London Business School who was born and brought up in Cairo. In the Arab Spring he went back to Cairo to his family to be there for them and everyone in Tahrir Square. This was his story. Twenty minutes alone on stage at Bloomberg in London in May 2011. He finished, 'We must all be leaders now.'

Lesson: Some stories just need to be simple, clear and authentic. There was no spin. Just a spine-tingling sense of 'I was there.'

Kevin Eyres

European MD of LinkedIn, spoke first at the London TED Bloomberg event. He was good but there were 22 speakers after him. Some were more charismatic but weeks later his clarity of messages came back again. You have to look after your employment brand; every job is a stepping stone; his induction speech to new recruits: 'Welcome to one of the best companies in the world ... we are going to transform you.'

Lesson: First impressions matter but it's content that ultimately wins. Presentations are about that 'blink' moment but great presenters also sow seeds.

 **brilliant** tip

Show passion even if you leash it in to fit your personality.

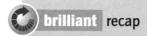 **brilliant** recap

In all my coaching I find the conversation always comes back to content and the shaping and sharpening of what is said.

There are style no-no's to be sure, like being uncomfortable or timid or incoherent. What you wear matters only insofar as what you can carry off – Steve Jobs is no style icon, but it works for him. How you sound matters because it has to be heard and understood. Voice matters and the best presenters have all had a lot of practice.

But looking at all my favourite presenters we see they:

● had something worth saying;

● said it engagingly and wrapped their audience in their story;

● were themselves enlarged to fill that stage but were not an impersonation of someone else;

- used personal anecdotes, which were very powerful;
- won the audience over with their expertise and depth of knowledge;
- were nice and liked being up there;
- kept it simple;
- enjoyed their slides and were playful with them;
- showed lots of energy.

Watch TED, go to conferences, become a student of speakers in all fields – stand-up comedy, newscasters, gurus. You'll come to understand how powerful an impact good presenting can have.

PART 5

How to become brilliant – pulling it all together

There are two ways of achieving the kind of rapid improvement anyone who's bought this book will want:

1 Practising in two ways:
   - in theory – by assimilating and then working on the lessons contained in this book;
   - in practice – by doing as many presentations as possible, big or small, to put those lessons into action.

2 Studying lots of presenters so you can to learn (good or bad) what the real secrets of success are on the stage as opposed to on the page.

One day we'll all be brilliant

When new schools starting up today state their primary objective as teaching command of the English language with, critically, an appetite to read, a fluency in writing and a confidence and competence in oral communication, you know that the presentational world is going to change.

Presenting well is a key business skill. It should be on every MBA course, it should be the first item on any graduate training programme. Bad habits get embedded the older an executive gets and those who are senior and are untrained presenters have been ill-served by their bosses in the past.

It is not just a nice-to-have personal skill but as basic as being able to write a decent document or read a balance sheet. Increasingly people are the key competitive differentiator in business. How people think, how they create things working together and how they manage the work streams which make great processes are what makes a difference. But for people and teams to succeed they need to be great communicators and sharers of ideas.

Why you must aim for brilliance

At its core, presentation is about the skill of communicating ideas. Ideas are what drive and change business so being unable to present well is to be deficient in one of the key business skills. To be a brilliant presenter is to have the potential to be a big business success.

You'll achieve brilliance through a desire to improve, constant practice, hard work on preparation and deep thought about content. The more you work at it the better you'll get. And the better you get the more you'll be asked to do it.

CHAPTER 13

Moving from good to brilliant

By now you'll have made a few discoveries about yourself and about some of the techniques you need to enable you to be brilliant. You should also be discovering that you are doing better than you'd ever thought and that, with just a bit more effort, you could be even better.

Until now presentation has not been in pole position for you but now it's time for a change. Because becoming a brilliant presenter is the most important, career-defining thing you can do. This being so, do it.

A change and a commitment

I want you to make me a promise – on your next presentation I want you to triple the amount of time you spend on it. Before you get too far into it, I want you to score yourself on each of the following on a scale 0–5, where 5 represents 'brilliant' and 0 equals 'appalling':

- context;
- story;
- colour;
- illustration;
- performance.

Scoring anything over 18 will mean you are approaching the 'brilliance zone'.

You will discover that you will achieve an instant improvement just by focusing on this framework. But the real keys now lie with the content. If you know that you have a compelling story you will feel more confident and more inclined to go that extra mile. So I want you to spend a long time (through the night if needs be) on the content – the story and the colours you want to introduce to that story.

The good news is you'll be astounded by the level of improvement you achieve. The bad news is you'll be very tired – and if you aren't tired you aren't working hard enough at it.

Getting the story right will drive everything else

This is not going to be easy because you are probably straying into unfamiliar territory, where the idea of building a brilliantly compelling story is alien to the way you work. What I'm asking for is something that is:

- descriptive and vivid;
- informative;
- attention-grabbing.

And this requires a lot of hard work.

Take a piece of paper and start making notes – anything that comes into your brain. Circle words that seem as though they might make themes; you know, the kind of words around which others cluster. Words like 'change', 'new people', 'disruption', 'simplicity', 'creativity', 'breakthrough', 'win' … these and many more. Cluster them and look for connections that apply in your story.

brilliant tip

Real brilliance kicks in when you start making connections between themes.

How do brilliant presenters do it?

They generally apply one of four approaches that you will recognise when you see them at work. These may also work for you.

- **The power of intelligence**: This is the application of considered thinking. People like academics, senior civil servants or experts in their field use this to great effect.

- **The presenter as commentator**: This is where irony is used or where the speaker sees things from several points of view and comments on the viewpoints. This is a common source of deadpan humour.

- **The power of modesty**: Or 'I am very clever but I don't take myself too seriously.' There is nothing better than seeing the big boss acting like a human being, or an expert speaking to non-experts at their level without being patronising. Not being too serious is a great way of getting very serious points across.

- **The power of knowledge and experience**: Deep knowledge lets you be very simple and very certain. Simon Walker, ex-chief executive of the British Private Equity and Venture Capital Association, talks of people who casually allude to something while appearing to have pocketfuls of knowledge about it should they choose to reveal them. It's always great to hear an expert talk about their stuff, whether a sportsman or Bill Clinton. In your case talk about what you really know and it will have the same effect.

The power of simplicity and brevity

As a simple rule, this alone will drive you to brilliance most compellingly. It was well recognised, not least by Mark Twain, that it is harder to write good short pieces than good long pieces. Stream of consciousness is easy but sonnets or haiku are harder to write than ballads.

Reducing your presentation to its perfect state takes effort and time. Too many presentations are, as it were, undercooked – the sauce is thin and runny, the ingredients haven't blended together. When you are pulling your presentation together have a blue pencil beside you which is your 'brilliance editor'.

I heard recently yet another great reason for using cryptic charts from Ian Parker, ex-CEO of one of the key divisions at Zurich Financial Services: 'Do not trap yourself beneath a welter of words on a slide. How embarrassing to have "Sales are continuing to grow fast" when you have just had a sudden embarrassing sales hiccough. The single word "growth" allows you flexibility.' He's right. I suppose you could manage to say 'Growth has been great but can it continue? Look at last month when we had poor sales. Yet the underlying story is still good if we can get our sales momentum back. It also serves to show that you can take nothing for granted and that you can never take your eye off the ball.'

 **tip**

The flexibility to respond to events as they occur is what modern management must have and modern presenters show.

Too many people inadvertently treat their presentations as suicide notes. Stories of the day before yesterday and worse … Their ill-digested, uncooked ramblings on screen may help them keep on track but pity the poor audience. The next time you review a presentation that you are going to give make sure you ask: 'How can I reduce the stuff in here? How can I take words out of the slides? How can I achieve clear focus on the issues?' It's great advice to ensure you are cryptic and that you aim for less because, as we know, less is more.

The better you get the more your curiosity grows

Be incurably curious. The stuff of all brilliant presentations is the unusual, the surprising and the novel. It's also true that people who have a big appetite for life and an unquenchable interest in it are generally the most entertaining company. Great presenters describe things vividly with the excitement of someone seeing something for the first time. They have a childlike sense of curiosity and an appetite for discovery, so try to be like them.

The best example I have of this was at the Hay-on-Wye book festival. Germaine Greer was running a master class on Blake's poem 'The Sick Rose'. The poem is only eight lines long and she spent over an hour dissecting it. This was one of the best hours I've spent for a long time. This now mature Australian sat on a chair in the centre of the stage and talked fast and intensely. She demonstrated fierce curiosity. She really wanted to know what the poem meant. She interrogated it, rained questions down on it and was a restless source of 'don't get it, don't know, don't know yet'. She wrestled with the words and pulled each one to the ground where she pummelled the meaning out of it. It was completely absorbing and totally brilliant.

Brilliance comes from talking and listening

Brilliance lies in pubs and shops, not at desks. This is a serious point. People working with other people will, on balance, be more likely to produce brilliant stuff than people flying solo. Brilliance is found outside yourself, not inside. What you do is provide the electricity that links the elements together.

Tom Peters said 'No one ever wasted money travelling' and he's right, insofar as the reality of getting out from behind your desk and seeing new things can make you do and think more brilliantly. Heathrow will inspire you. It doesn't? *It doesn't?*

Well it should. Heathrow *will* inspire you – a vast city of transit and shops, especially Terminal Five. Yes, truly, all human life is there. It's a market researcher's delight. Spend time listening, looking, tasting, touching and smelling. You'll learn more about human nature in the perfumerie in Selfridges, Harvey Nichols or Harrods than you ever will in an office.

You'll also learn more having a good lunch with an engaging friend than you will reading *The Times*, which should take only 20 minutes to do anyway; hopefully lunch will take longer. Brilliance comes from listening, absorbing and then recreating. Brilliance doesn't exist in an intellectual monastery or nunnery.

 **tip**

If you're stuck on a presentation, have lunch with someone interesting and opinionated.

Brilliance comes through learning

Work with professionals, work with the best and work with good teachers. Because brilliance will only come from pushing yourself and giving yourself frightening challenges. A challenge which evokes 'I really honestly couldn't do that, it's just beyond me – honestly', but which when met lifts you to new heights.

One of the people I coached seemed unlikely ever to be even a competent presenter. I'd now describe them as close to brilliant. The lesson to be learned is that you will find that you achieve brilliance quicker by working with coaches.

Steve Jobs learned and, today, so do the best and smartest executives. There isn't a brilliant anything, singer, golfer, athlete, actor, writer or whatever, who doesn't use the experience and inspiration of a good teacher to help them learn.

brilliant tips

Taking you into the brilliant zone

● **Taking the risk of being brilliant:** When doing a presentation, our instincts are to retreat into mediocrity. A 'brave presentation' has the same sound to it as the immortal 'That is a courageous decision, minister' in the BBC series *Yes Minister* (decoded this means it is a rash and possibly career-damaging decision). Yet it is impossible to be a brilliant presenter if you only ever play safe. Rupert Howell, MD of the commercial arm of ITV, was known to sing at the start of a presentation when he was in advertising. Bartholomew Sayle, who ran the Breakthrough Group, did the same to prove how far out on a limb he'd be prepared to put himself. I did it in a wedding speech recently. That's taking risks – and if you heard my voice you'd realise this is very high risk indeed. Who knows what Sebastian Coe did at the presentation of the London Olympics bid but we can be sure it wasn't ordinary. My brother-in-law used to say 'boldness be my friend' before playing a slightly risky shot in golf – and, surprisingly, it often is.

● **Steal brilliant ideas:** It was Picasso who said 'amateurs borrow, professionals steal' and if you want to be brilliant you must, above all, be professional. The other line I like is that if you borrow one thought it's called stealing, but if you borrow lots it's called research. So get your swag bag out and get stealing.

For a start, watch the best stand-up comedians and see how they work. Watch Jack Dee and see how he became 'king of laid back'. This happened because he failed as an enthusiastic comic. In the last few performances before finally jacking it in, he stopped really trying and adopted the characteristic Jack 'sod-it' Dee mode. The audience response was ecstatic, and the rest is history. Watch my namesake Rich Hall for controlled anger and for going off on an amazing rant about accountants

▶

or whatever seizes his mind that night. Watch Jo Brand to see the mistress of measured exposition and irony. Watch Rowan Atkinson to see submerged mania threatening you and a focused delivery where every word is like a sharp knife. On a more serious note (but not much more serious) watch Andrew Neill on the BBC's *This Week* to see a presentational tour de force being exploited by someone who uses every trick available.

Take every chance you can to see professionals at work then nick their best stuff, techniques and attitudes that seem to work for you – be shameless.

● **Being very brilliant means being 'very something ...':** The danger of being too reserved lies in simple human nature – in wanting to disappear into the crowd. As our nerves get worse we want to hide under the blanket, sucking our thumb. But we need to be very brave to be outstanding. Here are some options:

   – **Be very learned:** This works brilliantly if you are very learned and if you have the material to carry it off.

   – **Be very funny:** This is all about risk and reward. Clint Eastwood as Dirty Harry said 'Do you feel lucky, punk? Well do you?' I feel much the same about humour in business. It is best avoided unless you are very sure of yourself and even then. . .

   – **Be very theatrical:** This can work brilliantly depending on the personality of the presenter and the nature of the audience.

   – **Be very confident:** This will always work provided 'confident' doesn't become 'cocky'. Confident presenters calm an audience's fears and hostility; very confident presenters inspire them.

   – **Be very smart:** It is brilliant to be presented to by someone who convinces you that they really understand what's going on in a consumer's mind. George Davis, creator of the Per Una fashion range for Marks & Spencer, was convincing in his claim that he really, really understood women and what they wanted. He always sounded smart and on the ball.

- **Be very cross:** I don't recommend this unless you are Tom Peters who rants and raves and gets paid a fortune for doing it.

- **Be very enthusiastic:** Far too many managers in business seem worn down by their bosses or by the rigours of the job to the point that they seem unenthusiastic, weary and bored. Enthusiasm in presentations can lift you into the 'brilliance zone' faster than anything else I can think of.

- **Trust in yourself:** As I was lining up to play a tee shot over a lake, someone said to me, 'Trust in your swing.' Only years after that inevitable splash from a topped drive did I really work out quite what lay behind this. If you don't believe in the equipment and talent you have then you are in a very bad place. Be the first to praise yourself – be the first to say how great the good bits were and be confident enough to criticise the poorer bits.

There is one piece of advice I want you to copy and put in your pocket if your quest to be a brilliant presenter is genuine:

You are a brilliant presenter. Be yourself. Be active. Be fun. Every presentation you do is a 'crunch'. So just do it freshly, enthusiastically and with rigour. You can be a brilliant presenter. But start by feeling brilliant.

- **Write yourself a motivating letter:** This is an extension of what appears above and a tool with quite amazing impact. Brilliance can only happen in an environment where praise is bestowed. Apart from painters like Cezanne who had a rotten time in their life, most great artists and, by definition, most great performers, require praise to deliver the goods. So help yourself. Write that letter – 'Dear Richard, I want you to know you were great yesterday – really inspiring and funny but, more to the point, totally focused … it was great to be there … you were brilliant.' Self-indulgent? Ridiculous? Unnecessary? No – not if it achieves the key aim of making you improve and become, hopefully, brilliant.

brilliant recap

So much of what happens to us and so much to do with how we perform depends not on what we know but on how we feel. In the journey from ordinary to brilliant the biggest change will occur to our psyche rather than just our command of technique.

Daryll Scott, Creative Director at the interesting and highly successful company Noggin, says: 'Be prepared for anything rather than prepared for everything.' The latter is just tiring the former makes you very alert.

But in this quest for brilliance most of all be in a good mood. It's really hard being brilliant when you are grumpy. One of the reasons that so many politicians become less effective as they spend more time in office is that they get worn down by intransigent issues. They become fed up, irritable and frustrated and their one-time brilliance is blunted.

So the next time you start to prepare a presentation, and especially as you actually perform, think hard about how happy you are and how glad you are to be doing it, especially to this particular audience whom you particularly like. In doing so you'll discover a brilliant difference in the way you perform.

CHAPTER 14

So you've made it – brilliant, well done!

B ut can you remember how to do brilliant presentations time after time? Are you confident that the 'wow' of day one isn't a drab drizzle of a presenter on day two? Do not join the hero-to-zero-club, read this chapter before every presentation. However brilliant you are. However good everyone said the last presentation was. Take nothing for granted. Read it again and again. It'll stop you making sloppy mistakes or falling back into bad habits. It'll stop you getting cocky. As Robert Fitzsimmons the boxer said, 'The bigger they come the harder they fall.' You are at your most vulnerable when you are at your most brilliant. Don't stumble; your key activity is practice, practice and yet more practice.

This is the brilliant formula

There's nothing magic about this, you just need to listen to the professionals. Follow the 'brilliant presentation' format and work really hard at it. And remember, the idea of a quick fix is appealing but is as misguided in its effect as a crash diet:

- To start with you need to be honest about how good or bad you are.
- You need to practise really hard to get good.
- You need to devote more time to planning, writing and rehearsing than you could imagine possible.
- The biggest change that you'll notice is that the more you practise the more you will begin to *want* to present and,

however much you may protest, a certain thrill will fire in your stomach as opposed to a dread of anticipation. If you've read this book and are prepared to devote time to improvement then it is likely that you'll have improved massively already.

This is a very competitive world and there are many very young people who find the act of standing up in front of their peers a lot less alien than it used to be for those a generation older than them. In modern life, do not underestimate how important and career-defining being a brilliant presenter is. People are judged on how they perform in public because it shows:

- how good the company they represent is – the better they are, the better their company seems and the better they seem for representing it;
- how confident they are in their own ability;
- how well prepared they are;
- how knowledgeable they are;
- how attuned they are to their stakeholders;
- how inspirational they are as leaders or potential leaders;
- how responsive and flexible they seem to be;
- how with-it they are.

How with-it are you?

In a world shaking up and shaking down as much as ours is, a world in which the 'American dream' declines in the face of the 'Asian explosion', we'd better have a global viewpoint and a sense of why what *The Economist* mischievously called 'Chindia' is so important. These are not just 'emerging economies', they are the future of the world.

Our ability to deal with the new icon of the twenty first century, St Paradox, and communicate the challenges that exist to those around us in a clear and inspiring way will be what distinguishes the best from the average executive.

In both India and China I've found the concept of the 'presentation' still a little novel but increasingly becoming normal. A well put together and thought out presentation is good business and it's also good manners.

If you want to give a presentation some oriental spice, you'll struggle to find the word presentation in any book on China. But if you read Laurence Brahm's book *When Yes Means No (or Yes or Maybe)* you'll get a good grasp of 36 key Chinese business strategies. It's a great read.

Here are three to be going on with:

1 Kill with a borrowed knife – i.e. make use of someone else's resources to do your job.

2 Retreat is the best option – i.e. do not play the game your competitor wants you to play.

3 Turn yourself into being a host from being a guest – i.e. reverse your position to salvage a situation.

If these don't begin to give you a clue that presenting in China might not be entirely simple that would be a surprise.

Communicating change vividly, with compassion and real understanding, will become one of the most vital assets any executive can possess. To keep up with the latest issues and launches check out Fast Company (**www.fastcompany.com**) and Springwise (**www.springwise.com**) on a constant basis.

Presentation has never been more important

It's more important in business generally and to you specifically.

First of all to you. You've decided it matters to you and spread the ground bait showing you are more than an ordinary presenter. You have quite simply raised expectations. If, like anyone half competitive, you have raised your game there's only one way and that's up. You are no longer (to use that golf expression) a weekend golfer.

Second of all in business, brilliant presentation to stakeholders – workforce, peers, competitors, customers, consumers, suppliers, investors, media, analysts, business experts and opinion formers – has never been so important because it sells ideas, changes minds and makes business move forward. Brilliant presentation comes, more than anything else, from clarity of thinking and empathy with the audience, and the clear thinking and understanding that will give us the best chance of building a better world. Yes, that important. Communication has become the number one skill.

The tools of the discipline have been described already – but, as any brilliant presenter will tell you, repetition does not go amiss.

- Decide how good you are. Analyse your strengths and weaknesses. Given how important the art and craft of presentation are, do not try to wriggle away from your shortcomings. Talk to others even if you think you are perfect because their comments may be helpful. Get a fix on where you stand and what a reasonable improvement target is. This is about planning your future, not just hoping for the best.

- Contextualise your presentation. Spend as long on this as it takes before you even start thinking about what you are going to say and how you are going to say it. Work out *why* you are doing the presentation, to *whom*, how they *feel* and what they *think*, *what's* going on around them, *where* you are doing it, *how many* will be there. Leave nothing to chance – be very clear about the context in which you are performing. Get your understanding of context wrong and the rest will unravel.

> ### ✴ brilliant tip
>
> Leave nothing to chance – be prepared for anything.

- Refine the story. The message, the big idea. How does it develop? How can you tell it clearly and compellingly? Do you have the 'elevator pitch' or synopsis of the plot absolutely clear – so were someone to say 'Richard, you have only two minutes to do your presentation to us, not half an hour' could you do it? Become a great storyteller and you could become a great leader – it's that simple.

- Add splashes of colour. The stuff that enlivens, dramatises and makes it all exciting. The bits of contemporary fact, anecdotes, data, evidence that makes the story unforgettable and convincing. These are the spear carriers, the crowds, the extras, the props that make the story more fun. Remember that stage direction in Shakespeare's *The Winter's Tale*, 'Exit, pursued by a bear' – great stage direction, great fun. But it's more than fun we're talking about. It's about character, attitude and intelligence and they all add colour and memorability to a presentation.

- Get the visual backup absolutely right. The slides, the staging, the 'toys' you use to make a memorable point. These are there to enhance but never to run the show. Make sure they are good enough so they don't hamper the rest. Learn to do your own by all means but take professional advice when it comes to a big show so that you are up with or ahead of other speakers. Good slides speed you up, make you feel good and keep the audience on your side without demanding too much of them. Be proud of your material; don't accept that it's just good enough, or it almost certainly won't be good enough.

brilliant tip

Do your own slides by all means but when it comes to a big show take professional advice.

- Give a great performance. A nerveless tour de force is what you should be aiming for. Tame those butterflies in your stomach, make sure you are in good voice, use coaches to transform you from being apologetic (very bad) into a good-humoured energy source (very good). You are on stage – act like you own the space, the story and, for so long as you are up there, as though you single-handedly own the attention of the audience. Have fun. This is not a board meeting, this is theatre.

- Be in control. You are a control freak or, better still, you are surrounded by control freaks who'll leave nothing to chance. You have created a brilliantly unusual agenda and you have great takeaways. You are thinking of ways to break the 'me up there and you down here' paradigm. This is your show – go for it. Before the show, during it and afterwards. And if you do all this, this audience will be yours forever.

brilliant tip

Be a good-humoured energy source who owns that stage.

 recap

Brilliance will be achieved if … you do all the following. They all matter but tattoo the first five on your brain.

- **You believe life is exciting.**
- **You decide you really want to enjoy presenting, not just to go through the motions.**
- **You want to be great.**
- **You believe you can do it.**
- **You practise.**
- You succeed in controlling your nerves.
- You have an avid curiosity about everything you read and see.
- You listen as well as you talk.
- You realise what other people can do better than you.
- You use their superior skills when it matters.
- (So important I'm repeating it) you practise.

While I can give you techniques to succeed, I can't give you that desire to excel. But if your desire to be brilliant is obsessive then I think you are going to find that you like presenting a lot pretty soon.

Welcome to the world of obsession. Welcome to the theatre of presentation. Welcome to success in business. Welcome to brilliance. Enjoy the buzz.

CHAPTER 15

Applying your skills to all kinds of meetings

When I first wrote *Brilliant Presentation* I hoped it contained some sound advice and maybe some inspiring tips. Perhaps it did, because I've had positive feedback from a lot of people. However, some told me there was an unanswered question. When is a presentation not a presentation? What presentational behaviour is required in a small but crucial meeting? This is my guidance on how to run a meeting of 10 or less possibly using paper as a prompt but almost certainly using no slides.

No slides. Why not?

There are times when slides seem wrong. When they suggest that you are not open minded at all; that your mind is made up; that it's locked shut and that you are there just to sell them your point of view. An informal meeting, hopefully of minds, becomes or can become instead, a formal, unproductive or, worse, an antagonistic monologue. I've seen people thinking they were prepared but because of the paraphernalia they bring with them they lose the audience; as Tommy Cooper, late genius-comedian and improviser, used to say – 'Just like that.' Small meetings are designed to make things happen, not to be slide shows. The catchphrase 'and here is one I prepared earlier' is to be avoided. Instead you should be aiming for 'and here is one we've created, together, now'.

You've been promoted, so make that first meeting count

Decide what you want them to think about you and about it – your promotion. Without knowing the context of your promotion it's really hard to be helpful. Did your predecessor die of natural causes, commit suicide or were they executed – fired that is, or were they murdered – whistle-blown by colleagues or, alternatively, did they go on to higher and grander things? Context is everything. But this is your chance to make an impact so make it.

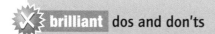 **brilliant** dos and don'ts

Don't

✗ Be flash.

✗ Show off.

✗ Patronise.

Enough of the do-nots.

Do

✔ Set a clear agenda.

✔ Aim to make decisions.

✔ Try to make the atmosphere relaxed, positive and effective.

✔ Keep it moving along fast.

✔ Try to make them walk out muttering 'That's more like it.'

You want to be promoted, so lift your game

The biggest reason for executives failing is because they are poor presenters. Work for hours on your own presentation to reduce an argument to five clearly argued minutes. Know your stuff so well you cannot be thrown off your stride. But be kind to yourself – everything falls into threes – three things to comment on, three problems, three opportunities. The 'rule of three' is for you a way of framing arguments and focusing on messages so don't wreck it by continually saying 'three things occur to me' and revealing that you rely on this technique (anyway they might have read this book). The reason for them promoting you is that you look as though you know what you are doing. The easiest way of making them think this is to be accomplished at talking about what you've done and are doing and why you are doing it.

Be prepared: the key advantage

I was never a boy scout. However, I always thought their slogan was spot on – 'be prepared'. All the people I coach get hammered by me on the issue of preparation. Astonishingly, many people prefer to wing it and this has always seemed to me the utmost in futile hope. You cannot seriously expect the god of presentation to keep on getting you out of the unprepared mess you've created or have just let happen. When all it takes is an hour (or a lot less if you get the hang of it) to get into some sort of shape, not preparing properly is a scandal.

Simply sit down and work out what your audience and colleagues want or expect. Look at all the angles. Read the papers. Think of the questions that need asking and the answers that need giving. Get yourself into a position where the meeting and anything it throws up is not a surprise.

Make your meeting sound as if it's worth going to

Making the meeting sound worth attending is an obvious but effective way of operating. Send out emails to tell people how it's going to work. Seek out colleagues ahead of the meeting so that unnecessary confrontation is avoided. Make the meeting a highlight in a continuum of discussions. If you have to prepare papers in advance, make sure they get to people in good time and that each topic has a simple one-page summary to allow people to get to the nub of things quickly.

Location: choose it, use it, make it memorable

If the meeting isn't in your office or a meeting room, see if you can find somewhere interesting to hold it.

● Outside – if it's a beautiful day, why not?

● On the river – if you are near one.

● Somewhere with an inspiring view.

● The War Rooms at Whitehall (if you want to create that sort of drama).

● A room which you've art directed ahead of the event with your products or a wall chart showing all the stakeholders – anything that focuses people on to an issue.

● Somewhere no one expects to meet but can't get away from. Imagine buying a capsule on the London Eye and holding a half-hour meeting there. Awesome above-the-battle stuff.

● A hotel suite – expensive but being there makes a statement that this meeting matters.

Define everyone's expectations. Meet them

Decide exactly why you are holding a meeting, what you want to get out of it, what others want to get out of it, spend 10 minutes on working out what the mindset of everyone coming to the meeting is. Imagine what the perfect meeting would be like from everyone's perspective. That's what you should aim for. Always try to meet expectations. It's easy to do. Ask everyone individually what their expectations are ahead of the meeting. Check out that you've met them at the end of the presentation.

Learn to think on your feet

It's easier to do than you'd think, yet before you acquire it as a technique it seems an unreachably distant aspiration. There are three parts to it:

1   Having mastery of your brief and the thrust of it, not just the (often) distracting detail.

2   Knowing the people and the way they behave. In some companies, challenge is a way of life.

3   Establishing a presence as a performer. Be consistent in your behaviour, be polite and establish the way you work and the way you respond. Think about how you want to be seen and heard.

Think about two things – one is being very light on your feet. What is required is nimbleness not dogma. The other is having a clear set of things you believe in; lessons you've learned; principles on which you wouldn't concede. Seeing a performer on stage or in a meeting who is clear about what they stand for is always impressive.

The executive summary

As people become more senior in business their attention span tends to shorten. As all of us become busier and busier our ability to process information reduces. We all suffer from overload. This is where the executive summary comes into its own.

Those brilliant at it will flourish and get on. And it's easy to do.

- One sentence summarising the issue: simple language, no jargon.
- How radical a change is involved?
- What do we need to consider?
- Who in the organisation does this affect?
- How does it affect them?
- Why is it important, critical, fundamental (whatever it is)?
- When do decisions have to be made and implementation started and completed?

- What are the cost implications (investment, pay back, cost saving, avoidance of risk)?
- Who has to do what and how do they have to do it?
- Summary of the above repeated.

Brand your meetings so they want to come again

How do you brand a meeting? Well, you could have a chocolate on each agenda or, better, a bowl of chopped carrots – for energy. You could have flowers on the table. Pansies would be good. This is what Ophelia said in *Hamlet*: 'And there is pansies, that's for thoughts.'

You could have something that symbolises the intent of the meeting. For a cost-cutting meeting you could have a knife (I'm joking). For growth you could have a marrow. For creative brainstorming you could have piles of Smarties or unusual products on the table. For cutting through red tape you could have scissors and red tape or red bureaucracy balloons with pins so you can burst them.

It's very simple. Tangible memory hooks make meetings swing.

Do great minutes

Do them fast and do them brief. Most of all, make sure they really reflect the meeting you had, not the meeting you wished you'd had. Stuff like this is all part of the presentation experience.

Have great coffee and backup material

Mark Weinberg appointed FCO as the advertising agency for Allied Hambro (as it was then called), on the strength of its coffee. Things like coffee, fresh orange juice, good biscuits all matter. They reflect how much you care. Presentation is about more than PowerPoint, it's about creating a total experience.

Pre-meetings and post-meetings: how to play the game

When we realise that we are creating a situation that could be called 'presentation for results' we realise it isn't an exercise in presentation craft so much as office politics. The presentation is merely the medium whereby you get the results you want. So make sure you prime people, make sure they are properly briefed and that you follow up with them afterwards.

The game is to move things on. The game is to involve everyone so you get a more intense and productive response. Make sure you make decisions, agree actions and that the participants at this presentation/meeting realise they've been treated seriously.

 **recap**

Meetings should be and can be well crafted events in which the mood is such that you achieve:

● alignment;

● improved thinking;

● a shared mission to communicate more powerfully.

Everything that distinguishes a brilliant presenter on stage can be tuned down to apply to running great meetings that have presentation content. Not only do they work better but they are more fun too.

So enjoy yourself and make sure everyone else enjoys themselves too.

CHAPTER 16

Summary of the lessons in this book

By now you'll be an expert in modern presenting. You'll know how to present, what to avoid and ways of making yourself stand out. In this world of change where the impossible happening begins to seem normal and where we begin to believe that anyone can achieve their goals if they are determined enough, it's important to keep things in perspective. because some things remain unalterable. A great orator of the past would adjust to be a great orator of today. Dickens would have been a great blogger. Shakespeare would have Tweeted if he'd lived today (and written things like the The Wire).

Some things never change ...

So this is a reminder that although things seem to be changing at an incredible rate, things like love, marriage, children, fear, greed, ambition, storytelling, resourcefulness and laughter all remain pretty well in place.

brilliant tip

There are basics in presentations which never change ... being audible, charming and energetic for starters.

... But others do

What has changed are five key things when it comes to the world
of presenting:

1   The ubiquity of the presentation – presentations are the
    norm now everywhere, which means people's expectations
    of them are rising dramatically. Suddenly the papers are full
    of events with people speaking. Conferences used to be very
    expensive one-offs, now they are reasonably well-priced
    and frequent. The most important knowledge exchange
    is happening on stage, live and not just in books and
    academic papers.

> **brilliant** tip
>
> You'll learn more being at good presentations than you will at your
> desk.

2   English is the 'lingua franca' when global companies
    meet, which means the need for clarity and simplicity of
    expression carries greater weight than ever. But it's hard
    not to be impressed by the fluency with which the smart
    young from China, Iran, Turkey, Greece, Italy, Brazil and,
    yes, France, transact and debate in English. (Every time I
    write 'English is the lingua franca' I guess it must really rile
    the French. But they speak English so well, most of them, I
    guess they don't care.)

> **brilliant** tip
>
> Learn to speak easy-to-understand, simple and clear English. Avoid
> complex ideas and vagueness.

3 Technological developments mean new opportunities –
   approaches to presenting are being created the whole time.
   Things like video or animation are relatively easy (at last).
   You can borrow bits of film to dramatise your presentation.
   But be careful – don't interrupt your message with a mess
   of technology.

brilliant tip

Use technology but don't be ruled by it.

4 Increasingly presentations are becoming conversations with
   a more relaxed style, more use of personal anecdote and
   more focus on one or two key messages.

brilliant tip

Think of your presentation as a friendly conversation – that's the
current vogue. See if it suits you.

5 Teaching people to think on their feet and be great
   presenters is starting at four years old now. Here's what
   Peter Hyman, Deputy Head Teacher, and leader of a new
   school in Newham, East London, and ex-political strategist,
   says of his school: 'English language will be the main
   specialism, with a mission that every child leaves with great
   communication skills: an avid reader, a fluent writer and a
   confident speaker.'

So presentation is becoming even more important, everywhere.

 **tip**

Confident speaking is top of the agenda. Don't wait to catch up with the trend.

The eternal truths

At the same time there are some eternal truths which relate to performance in which clarity of delivery, the use of confidence and of charm are still key virtues. As ever, people want to watch and listen to someone who seems to know their stuff and is happy to be up there talking.

But the key ingredient which has never changed and never will is the importance of engaging, revealing and surprising content. People want a great story, a presentation with a 'plot', not a list of figures or a lot of jargon.

brilliant **tip**

Audiences always want to be engaged, informed and entertained. This will never change.

Death to jargon is easier said than done (and everyone agrees it must go) because when push comes to shove, rather like getting rid of clichés, jargon seems hard to remove.

The biggest current force for good in getting back to good, old values of simple, clear and colourfully told stories is of course TED and the influence they have. The quality of their performers – clever and passionate people talking briefly about an idea they love and want to share – is completely compelling.

TED has helped make 'brilliant' a benchmark. That's why in listing the topics of TED talks they have a specific section for what they call 'jaw-dropping'.

A reminder of how to be a brilliant presenter

Being a jaw-droppingly brilliant presenter is about attitude of mind. You've got to want to do it. You've got to believe you can be brilliant.

The biggest challenge starts in the preparation phase. This is when the seeds of disaster can be sown unless you are diligent, careful and enthusiastic. So prepare carefully and leave nothing to chance. Always allow twice as much time in preparation as you'd normally expect. Brilliant presenters rarely busk it. They work at their presentation, polishing it, editing it and rehearsing it.

The three key tips are:

1   Be very clear why the presentation is being given, who will be in the audience, what their expectations are and how much they know about you.

2   Construct your presentation so it addresses a set of questions and gives clear and understandable answers to those questions.

3   Do not talk down to the audience or use jargon. Make sure people know what you are saying and why you are saying it and never let them think in a puzzled way as you finish speaking, 'What was that all about?'

brilliant dos and don'ts

Do

✔ Know what your audience expect.

✔ Love your audience.

✔ Relish your story.

✔ Keep it simple.

✔ Never over-run.

Don't

✗ Be woolly. Every assertion must have evidence … add the word 'because' after anything you claim.

✗ Tell people what they know. Or if you do, give it a twist.

✗ Rely on research alone. Use personal anecdotes.

✗ Mumble. Make sure they can hear every single word you say.

✗ Use long, unusual words. Speak modern, global English.

The biggest challenges to brilliance

● Having too little time to prepare.

● Being self-indulgent. Remember, 'less is more'.

● Being too nervous. You are not going to die up there and you might as well enjoy it.

● Being cocky. The worst presentations are when the presenter has no nerves at all and comes across as condescending.

 recap

My personal top five brilliant tips are as follows:

Brilliant tip 1

Make them hang on your every word: 'Once upon a time...' Yes, have a story worth telling. Your best chance of being a success is to be a good storyteller with ideas worth sharing and the ability to build suspense.

Brilliant tip 2

Memory hooks make the difference: If you want to be remembered, try to find some splashes of colour that make you stand out – a few great facts, a great quote or a topical piece of evidence or a controversial assertion. Be brave.

Brilliant tip 3

Learn to think on your feet: You need to be a note-free presenter if you want to compete at the highest level. This is impossible to do unless your story is simple, clear and one you believe in and with which you are comfortable.

Brilliant tip 4

Be professional and work with professionals: Get a coach to accelerate your improvement. Do not try to do your own slides – use a specialist firm or make sure your secretary or PA gets top-class training in being brilliant at slides. If it's an event that could influence your career, work with the best and with the top sound and vision operators. Never be guilty of being amateurish. Make sure you rehearse at the venue itself. Do not be taken by surprise.

Brilliant tip 5

What a great audience! Make a conscious decision to 'love' your audience. Try and give off positive vibes. If you can get on with the audience they'll pay closer attention to you. And yes, smile at them, charm them and try to inspire them. They are your friends.

the brilliant series

 9780273742524 — brilliant Business Plan — What to know and do to make the perfect plan

 9780273742555 — brilliant Email — How to win back time and increase your productivity

 9780273740544 — brilliant Influence — What the most influential people know, do and say

 9780273743231 — brilliant Manager — What the best managers know, do and say

 9780273762423 — brilliant Coaching — How to be a brilliant coach in your workplace

 9780273743248 — brilliant Negotiations — What the best negotiators know, do and say

 9780273737452 — brilliant Online Marketing — How to use the internet to market your business

 9780273744092 — brilliant Time Management — What the most productive people know, do and say

 9780273744740 — brilliant Teams — What to know, do and say to make a brilliant team

 9780273744634 — brilliant Freelancer — Discover the power of your own success

 9780273744580 — brilliant Business Writing — How to inspire, engage and persuade through words

 9780273722328 — brilliant Project Management — What the best project managers know, do and say

 9780273726463 — brilliant Selling — What the best salespeople know, do and say

 9780273725114 — brilliant Pitch — What to know, do and say to make the perfect pitch

 9780273743217 — brilliant Networking — What the best networkers know, do and say

 9780273721239 — brilliant Marketing — What the best marketers know, do and say

 9780273762461 — brilliant Presentations — What the best presenters know, do and say

 9780273727347 — brilliant Copywriting — How to craft the most interesting and effective copy imaginable

 9780273721826 — brilliant Meetings — What to know, do and say to have fewer, better meetings

 9780273738077 — brilliant Customer Service — What to know, do and say to keep your customers happy

Whatever your level, we'll get you to the next one.

Available to buy now online and from all good bookshops
www.pearson-books.com